Chirality and Life

Rolf M. Flügel

Chirality and Life

A Short Introduction to the Early Phases of Chemical Evolution

Springer

Rolf M. Flügel
Dr. Hermann-Brunnstrasse 43
69198 Schriesheim
Germany
r.m.fluegel@dkfz-heidelberg.de

ISBN 978-3-642-16976-2 e-ISBN 978-3-642-16977-9
DOI 10.1007/978-3-642-16977-9
Springer Heidelberg Dordrecht London New York

Cover design: WMXDesign GmbH, Heidelberg, Germany

Printed on acid-free paper

Springer is part of Springer Science+Business Media (www.springer.com)

Preface and Acknowledgments

This booklet deals with the molecular evolution before the evolution of species. One of my intentions for writing this book was to raise questions on molecular beginnings of life. The problem is that there are no simple answers. However, hopefully, the questions inspire and spark curiosity and passion for science and pleasure especially in younger readers. I feel it is imperative to show that the essence of science is the idea and the ensuing enquiry, and in addition, its thrill. It is exciting to read about the unknown and to discover and explore the stunning beauty in Nature's complexity. The intellectual pursuit of thc origin of life is more than fascinating fun.

According to Karl Popper, the method of science is the method of bold assumptions, of inventive and serious experiments to disprove. Karl Popper said that our knowledge based on hypotheses is assumptions. Knowledge of assumption has no final validity. For instance, Newton's theory of gravitation cannot explain the orbital properties of the planet Mercury; Einstein's theory, however, took account of them.

The book consists of several parts. The first is a short introduction to the guiding principles and building blocks of chemical evolution. It contains a chapter on the important aspects of chirality or handedness of biomolecules. The second part is an attempt to describe up-to-date most plausible hypotheses and visions of the ancient world scenarios. References and specific websites will be helpful for those who wish to know more on specific topics.

My friends encouraged and helped me to accomplish this book. Volker Kinzel encouraged me with his invaluable advice. Gholamreza Darai helped me with his unorthodox ideas and comments. I thank Erwin Heiser for his discussions, especially on astrophysical and geophysical issues, and his helpful comments and criticisms. I wish to express my special gratitude to Volker Erdmann, FU Berlin, and Frau Susanne Dathe of the Springer Verlag in Heidelberg for advice and support. All of them discussed it with me, critically read, and corrected the versions of my book. It is due to their encouragement, comments, and insightful suggestions that I learned to complete the story and to work personal experiences into the text. I thank those scientists who exchanged information, communicated with me by email, and responded to my questions. I thank Harald zur Hausen for permission to use the facilities of the Central Library and the Central Department of Electronic Data.

I thank Paul Schendel who helped me and improved the text in style and language. It is due to Paul's suggestions and critical but constructive ideas that the book became more readable.

Schriesheim, Baden-Württemberg, Germany Rolf M. Flügel
September 2010

Contents

Abbreviations

ATPase	The enzyme adenosine triphosphatase removes phosphate residues from adenosine-5′-phosphate ATP
CoA	Coenzyme A, a molecule that acts as cofactor for many enzymes
CPL	Circularly polarized light
ee	Enantiomeric excess
eV	Electron Volt
LUCA	Last universal common ancestor
mRNA	Messenger RNA
NAD	Nicotine adenine dinucleotide
ptm	Posttranslational modification
R	A residue in chemical formula
RNAse	The enzyme that cleaves RNA
rRNA	Ribosomal RNA
SRP	Signal recognition particle
TCA	Tricarbonic acid cycle
tRNA	Transfer RNA
UV	Ultraviolet

Part I
Physicochemical Principles

Chapter 1
Introduction

Some major physical principles serve to direct the evolution and interaction of molecules. For instance, most molecules carry information so that they interact with each other. Another principle is that evolution proceeds from the simple to the complex.

Darwin's object of natural selection is the complete genotype. His theory postulates an individual common ancestor of organisms. Selection is a principle derived from the law of exponential growth rate. Natural selection started to drive evolution as soon as molecular replication became possible and operates on randomly selected chemical biopolymers that confer fitness upon their hosts. Fitness concerns not only biomolecules and their contacts within a network of pathways inside the cells but also the consequent effect on the cells, organs, and organisms. Fitness depends on the environment, which can change on both a local and a global scale, even because of cosmic events, for instance, meteorite impacts. A model for fitness ought to capture most features of these events.

Another common theme of evolution is variability on all levels of molecular structure, function, and organization. RNAs, proteins, and even amino acids often have different and multiple functions that are context-dependent. For instance, proteins can act as biocatalysts accelerating a biochemical reaction specifically without consuming itself.

The notion of catalysis is important. There are many ways to catalyze any reaction. Catalysts are subject to evolution with a tendency to progress towards those that become more sophisticated. An especially relevant case is autocatalysis in which the reaction product catalyzes its own synthesis. In a similar vein, distinct RNA molecules can act as quasi enzymes called ribozymes. Two crucial features of life are replication and metabolism in liquid water.

Replication is a way for molecules, individual cells, or whole organisms to produce progeny. Metabolism is the way an organism processes energy and to carry on the business of being alive. Both features are required and intricately intertwined.

It is likely that replication and metabolism were invented first followed by the development of cells. However, the problem of the origin of the first cell remains unsolved.

R.M. Flügel, *Chirality and Life*, DOI 10.1007/978-3-642-16977-9_1,

Metabolism is the mechanism by which organisms and complexes of higher order seemingly circumvent the second law of thermodynamics, the law of increasing entropy. Living organisms, by necessity, are highly ordered and have low entropy.

The DNA molecule is a right-handed helical polymer composed of two strands. Specimens of DNA pulled into fibers fitted the observed X-ray diffraction patterns solving the structure and function of DNA. The sugar phosphates are outside. The bases are inside the helix and form complementary base pairs between the large purines (A and G) on one chain and a smaller pyrimidine base (T and C) on the other chain. Thus, base "A" pairs with T, whereas base G pairs with C. As a direct consequence of the base-pairing mechanism, it becomes obvious that DNA carries information by means of the linear sequence of its nucleotides.

Evolution reflects the history of Nature. The ability to evolve through the combined effects of natural selection and inherited variation is the unifying principle of biology. Physical laws and chemical reactions dominated Earth before and after life evolved. The laws of Nature are the prerequisites of life. Thus, life is a chemical system capable of undergoing Darwinian evolution. In analogy to the synthesis of water from hydrogen and oxygen, life represents the emergence of a completely sudden and unpredictable novelty! Evolution is a fact, a reality no intelligent individual can deny. Evolution acts on nearly all levels from the first molecules to the most recent ecosystems. She seems to be an extravagant woman full of surprises.

Evolution takes place continuously and it is contingent. Thus, its interconnections form in a complex way so that novel functions can evolve in older contexts. Major constraints of direct contingency are global catastrophes, for instance, impacts of large meteorites or asteroids, glaciations of large areas, major volcanic events, and earthquakes. All these events can result in mass extinction of organisms.

On the other hand, the impacts of meteorites and comets affected the availability of distinct, biologically essential molecules on Earth. For instance, glycerol, nucleobases, glycine, and alpha-methylated amino acids are among the proven compounds delivered to Earth. The astrophysical website http://www.cv.nrao.edu/~awootten/allmols.html is useful.

It is a fact that there are organisms that can exist under extreme conditions that are plainly hostile. For instance, radiation-resistant organisms can propagate without photosynthesis deep within the ocean, under very high pressure, and partially in symbiosis with bacteria. In addition, the range of temperature in which these organisms survive is extremely broad. The upper limit seems to be about 110°C and the lower limit is unknown but might be as low as −100°C. Evolution is dynamic, versatile, is not aimed at distinct purposes and not running according to a plan since it has no foresight. It is slow, opportunistic, innovative, unpredictable, and stochastic. Darwin's natural selection is the first instance of a statistical law of nature. Natural selection is the main mechanism responsible for adaptation of organisms to their environment. The internal aspect of evolution is the developmental trend towards hierarchical patterns of interactions in developmental pathways and the ways in which the internal co-adaptation of organisms is maintained and enhanced. Multiplicity of functions of different factors plays a decisive role during the course of evolution.

There are important, non-Darwinian processes, too. Horizontal gene transfer of genes from cell-to-cell confounds the analysis of rooted phylogenetic trees. So do other epigenetic processes, for instance, symbiosis, endosymbiosis, virus infections, and, last but not least, epigenetic programming. This involves investigation of heritable changes in gene function that occur without a change in the sequence of nuclear DNA

Both prokaryotic and eukaryotic endosymbiosis increased horizontal gene transfer. Symbiosis is an important evolutionary concept since it works as efficiently as an ongoing cooperation; it is a major driving force behind evolution.

Chemical evolution is the increase in numbers and complexity of the newly formed compounds and polymers. The exogenous and indigenous compounds do have similar structures and more importantly, the built-in potential to react with each other to form complex biomolecules. They can be tested and reproduced in the lab, However, the nature of the formed key molecules depended on the environment, and since the environmental factors depended on the physical conditions of the early stages of the Earth (temperature, UV-irradiation, pressure), consequently, the nature of the compounds changed during the early phases of the Earth. One should not wonder why the composition of the Earth's ancient atmosphere is controversial. It changed quite often.

Evolution opens new spaces to play and tinker with so that processes originated that had to fit to each other. Chemical reactions sufficiently stable yet with changing information and a containment separated "inside and outside." Evolution was possible by way of changeability of the information carrier.

Chapter 2
Formation of Primordial Bioorganic Molecules

Chemical evolution concerns the chemical processes that occurred on the ancient Earth about 4.5–3.5 million years ago. It is important to note that it preceded biological evolution that resulted in the formation of protocells. These forerunners of today's living cells were capable of self-reproduction at the expense of some protometabolism. After Oparin's ideas of the origin of life became widely known and especially after Stanley Miller reported his prebiotic soup experiments in 1953, the concept of chemical evolution became accepted.

Living matter consists of six indispensable elements, namely C, H, O, N, P, and S, and, in addition, many more trace elements, for instance, iron. These elements are capable of existing in different valence states and are thought to have predominated the ancient Earth.

Different sources of energy were available on the surface and in the upper mantle of the early Earth. The sources of energy include volcanism, sun light, electricity, natural radioactivity, cosmic radiation, and reaction enthalpies.

The main reaction products of prebiotic chemistry were H_2, H_2O, CH_4, CO, CO_2, NH_3, and N_2. These compounds formed many intermediates including ions and radicals. The more important molecules that formed were formaldehyde HCHO, hydrogen cyanide HCN, phosphate ions, and cyan amide NH_2CN. The final spectrum of products encompassed glycerol, glyceraldehyde – the parent compounds of sugars – carboxylic acids, amino acids, urea, guanidine, purines, and pyrimidines. As an example for the many possible interactions the formation of the nucleobase, uracil is shown in Fig. 2.1.

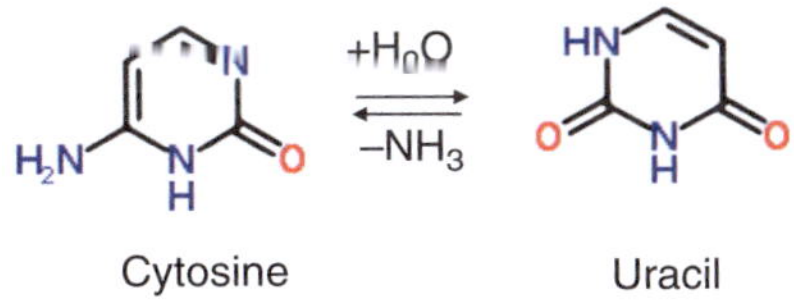

Fig. 2.1 Cytosine reacts with water to form uracil. The reverse reaction occurs when uracil reacts with ammonia

R.M. Flügel, *Chirality and Life*, DOI 10.1007/978-3-642-16977-9_2,

Chapter 3
Hidden Aspects of Chirality

Chirality or handedness is a characteristic feature of living systems. Natural L-amino acids and D-ribose exemplify the preference for one particular mirror image form, called dextrorotary (D) and levorotary (L) enantiomeric configuration. Thus, an asymmetric or chiral carbon atom is one around which four different substituents can be arranged in left- or right-handed orientation. A chiral compound is present in two mirror image molecules (enantiomers). Extant life depends on chiral homogeneity for the structure and function of biopolymers.

The natural standard amino acids are left-handed, whereas carbohydrates and nucleic acids are right-handed. This property has consequences: *the chiral groups arrange in particular patterns in space relative to each other*. Two differently arranged molecules that carry the same substituents are enantiomers. They differ in their ability to rotate plane-polarized light by equal amounts but in opposite directions. A mixture of equal parts of an optically active isomer and its enantiomer is a racemate and does not have a net rotation of plane-polarized light.

Figure 3.1 shows the two enantiomers of L- and D-leucine.

In addition, crystals show characteristic symmetry properties with respect to their forms and surfaces. Crystals that cannot superimpose themselves are chiral and occur in Nature, for instance, quartz and sodium chlorate. Louis Pasteur noticed that ammonium sodium tartrate came in two asymmetric forms that were mirror images of one another. Pasteur manually separated the left and right crystal shapes from each other to form two piles of crystals: in solution, one form rotated light to the left, the other to the right, while an equal mixture of the two forms canceled each other's effect and, does not rotate the polarization of light.

Life would have been impossible without the chirality of its molecules, which is required for the larger spectrum of biomolecules, their 3-dimensional assembly, and for the functions of biopolymers. In this balance, the unnatural, nonprotein amino acids detected in some meteorites may have played an especially significant role.

The question of the origin of homochirality is widely discussed but remains unproven. The proposal that peptide nucleic acid in which a peptide chain replaced the phosphodiester backbone had a role in the prebiotic origin of life prior to an RNA world has not been tested experimentally.

Does the way by which chirality originated tell us anything about life's origin? It is noteworthy that several routes exist that led to the formation of chiral compounds.

R.M. Flügel, *Chirality and Life*, DOI 10.1007/978-3-642-16977-9_3,

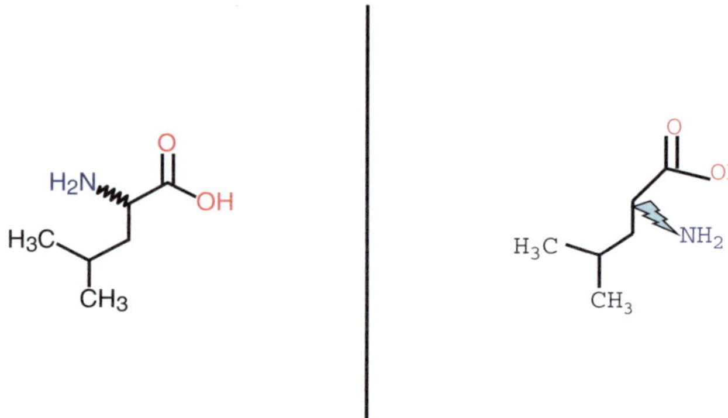

L-leucine D-leucine

Fig. 3.1 The middle line symbolizes a mirror. On the left hand, L-leucine is shown, on the right, its mirror image D-leucine. In aqueous solution, both isomers occur as zwitter ions in which the NH_2-group is positively charged, whereas the COOH group is negatively charged

This holds true for amino acids, nucleobases, and carbohydrates. Even within one class of the biogenic molecules, two or more independent routes developed. As illustrated by the formose reaction, chiral sugar molecules formed when catalyzed by an appropriate chiral amino acid (to be discussed further below).

One origin of chirality is a consequence of the different solubility of racemates and their corresponding enantiomers. Upon evaporation and, alternatively, by mechanical stirring in solution, they display different behaviors during crystallization and result in slight enantiomeric excess (ee). As described below in more detail, the natural L-amino acids likely determined the D-handedness of sugars.

3.1 Asymmetric UV Photolysis of a Racemic Amino Acid

Chiral sugar molecules may form by aldol condensations either with a catalyst such as proline or without any catalyst. In general, several routes exist that led to the formation of chiral compounds. This holds true for amino acids, nucleobases, and carbohydrates. Asymmetric ultraviolet (UV) photolysis of racemic D-and L-leucine in the solid state was performed in an attempt to simulate interstellar high vacuum and low-temperature processes (Flores et al. 1977). Leucine has the advantage that enantiomeric separation techniques provide a high separation factor, a high resolution, and racemization rate compared to other amino acids. The highest gain in ee was +2.6% after irradiation of leucine racemate by a synchrotron using circularly polarized UV-light (CPL) with 6.8 eV for 3 h. This setting simulates exposure to neutron stars. The data support the assumption that tiny interstellar ice grains can experience asymmetric reactions when irradiated by CPL UV-light. The result requires confirmation by using other amino acids.

3.2 Occurrences of α-Methyl Amino Acids in Meteorites

Sandra Pizzarello and her team were the first to publish that carbonaceous chondritic meteorites contain distinct α-methylated amino acids in low ee (Cronin and Pizzarello 1997; Pizzarello 2006; Pizzarello et al. 2008). As confirmed independently, the Murchison meteorite contained the following nonproteinogenic amino acids in ee: 15.2% L-isovaline, 2.8% L-α-methyl-norvaline, 2.8% L-α-methyl-valine, and 9.1% L-α-methyl-isoleucine (Cronin and Pizzarello 1997). As an example, Fig. 3.2 shows the structure of L-α-methyl-norvaline.

These four nonnatural amino acids are of extraterrestrial origin, since isotopic distributions of carbon and nitrogen are different from those of terrestrial amino acids. In addition, pristine samples of a meteorite found in South Pole regions collected under controlled conditions showed similar ee of α-methylated amino acids. The natural amino acids in the meteorites are racemic due to the ionization of their α-proton. In contrast, α-alkylated amino acids cannot completely racemize since they do not have a proton but a methyl group at the α-position. Terrestrial chirality likely originated from distinct meteorites since the α-methyl-amino acids of the Murchison meteorite occur in enantiomeric excesses. It is unknown whether the meteorites contained most of the standard amino acids during about 4.5 billion years the earth and the sun travelled through space.

In virtually all experiments that simulate the synthesis of a primordial soup, enantiomers of amino acids and sugars do not occur; instead only racemates have been produced (Miller 1953). It is difficult to imagine how only one of the enantiomers formed under the conditions of a primordial broth. Instead, import and identification of biologically relevant molecules in meteorites and comets appears as a more straightforward path. The compounds found in a meteorite provide a natural record of prebiotic chemistry in the early solar system and is closest to the onset of life.

The events that led to this ee may predate the origin of life on Earth. The Murchison meteorite contains almost all natural amino acids as racemates and, importantly, some additional anomalous amino acids that do not occur on Earth (Cronin and Pizzarello 1997; Pizzarello 2006). Among them are four amino acids that do not carry a proton on the asymmetric α-carbon atom but a methyl group. Among the four 'unnatural' amino acids that carry α-methyl substituents is L-α-methyl-allo-isoleucine. The systematic chemical name of the latter is 2-amino-2.3.-dimethyl-penta-carbonic acid, and it has two chiral centers. Therefore, it should

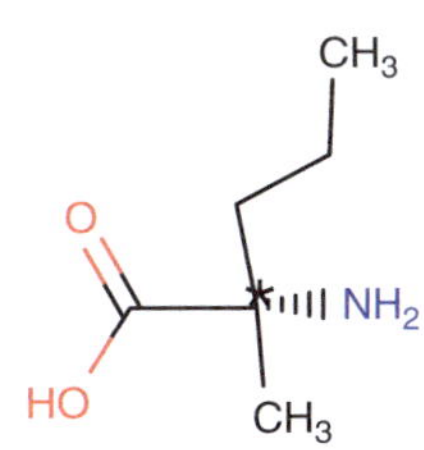

Fig. 3.2 The formula shows the structure of an alpha-methylated amino acid. The chiral C-atom is marked with an *asterisk*. It does not possess a proton in the alpha position

be possible to resolve it into four diastereoisomers. In fact, a research group showed this to be a case for the pristine meteorite GRA recovered undamaged from Antarctica (Pizzarello et al. 2006).

The L enantiomers of α- methyl-isoleucine and α-methyl-allo-isoleucine are in fact the four diastereomers of 2-amino-2.3-dimethyl-penta-carbonic acid. The crucial result of the analysis of these α-substituted amino acids showed that their L-enantiomers each had an ee of 12.1.4% and 14.0%, respectively, similar to Murchison's (Pizzarello et al. 2006).

Altogether, the number and range of the molecules in the Murchison meteorite were larger and broader than those obtained from prebiotic experiments. The compounds of the Murchison meteorite consisted of an abundant and diverse repertoire of organics that range in complexity from low molecular weight soluble compounds to insoluble macromolecular material (Pizzarello et al. 2008). The acid-insoluble material of the Murchison constitutes the larger portion (about 80%) of carbon-containing compounds.

The Murchison meteorite belongs to the type CM2, a carbonaceous chondrite. It was saved in large undamaged pieces immediately after fall-down without any terrestrial contamination near Murchison, Australia, on September 29, 1969, as documented by eyewitnesses. Its age was determined at about 4.4–4.6 billion years. It contains 12% water and surprisingly, almost about one hundred – precisely 96% – natural amino acids except for tyrosine, phenylalanine, and methionine, and a relatively high concentration of about 2 mM of HCN, NH_3, and CH_3CHO. Precise analysis revealed the decisive chiral stereochemical structure of the amino acids in the Murchison meteorite (Pizzarello 2006). It is likely that these amino acids once formed from the corresponding aldehydes by reacting with ammonia via Strecker synthesis. It is remarkable that in this particular meteorite sugars, pyrimidines, purines, and a high percentage of 29% water-soluble alkylphosphonic acid were detectable. Among them were some carbohydrate derivatives, erythrol, di-hydroxy-butyric acids, ribitol, glucitol, and, in addition, di-amino acids, malonic acid, and imino-dicarboxylic acids (Cooper et al. 2001). It is noteworthy that glycerol, the parent compound of the different membrane lipids of bacteria and archaea, was detectable (Pereto et al. 2004). The conclusion is that meteorites did not just deliver one class of several different compounds to the early Earth, indicating the potential importance of meteorites in the origin of biomolecules. A very similar spectrum of compounds and elements was detectable in another carbonaceous meteorite, the Murray meteorite. Although glycerol was present, glycerol-phosphates, however, were not present in meteorites.

In two independent studies that simulated comet chemistry, racemic amino acids formed after irradiation with UV light (Bernstein et al. 2002; Munoz-Caro et al. 2002). The results indicate that the distribution of amino acids correlated with the abundance of hydrous silicates. This raises the prospect of a participation of the primary asymmetric phases of minerals in the ee observed in the α-methyl-amino acids in the Murchison meteorite. This result points to a possible inorganic catalysis on the surface of minerals as a source of the ee of amino acids formed to various extent in interstellar and planetary abiotic processes.

Experiments showed that individual amino acid enantiomers from the Murchison meteorite are richer in the nitrogen isotope 15 relative to their terrestrial counterparts. This holds for L-alanine and L-glutamic acid (Engel and Macko 1997). The results unambiguously confirm an extraterrestrial source for an L-enantiomeric excess of distinct amino acids in our Solar System. In addition, the meteoritic amino acids were higher in the C-13 isotope, completely different from the terrestrial analogs, thus ruling out any terrestrial contamination. These completely different lines of experimental evidence are compelling (Pizzarello et al. 2006).

3.3 Partial Transfer of Enantiomeric Excess of α-Alkyl Amino Acids to Standard Amino Acids

Starting from the identification of chiral amino acids found in meteorites, a partial transfer to other biomolecules of low ee succeeded (Breslow et al. 2010 and references of Breslow cited therein). Under solvent-free conditions followed by evaporation and heating, D-α-methyl-valine reacted with pyruvate to form D-α-alanine with a relatively low ee of 3% by way of transamination. Higher enantioselectivity was observed with sodium phenyl-pyruvate. In this reaction, the D-α-methyl-valine plays two roles. It carries out the transamination that converts the keto-acid to an amino acid, while becoming a ketone after hydrolysis. Secondly, it is the source of a proton on the alpha carbon atom of the amino acid product, delivering it stereoselectively (Breslow et al. 2010).

3.4 Amplification of Enantiomeric Concentrations Without Catalysis

Chiral amplification of amino acids resulted in enantiomeric enrichment in experiments upon repeated evaporation carried out under prebiotic conditions (Breslow et al. 2010 and references cited therein). One origin of chirality is a consequence of the different solubilities of racemates and their pure corresponding enantiomers. Reports indicate that upon repeated evaporation of one enantiomer of an amino acid with a small ee led to an amplification of the ee concentration in solution. Hence, two crystallizations of an initial concentration of 1% L-phenylalanine in water yielded a final ee of 90.9%. This result is remarkable as it required no catalysis and occurred under credible prebiotic conditions (Breslow et al. 2010).

3.5 Thermodynamic Control of Asymmetric Amplification in Amino Acid Catalysis

Klussmann et al. (2006) found an alternative mechanism based on the equilibrium solid-liquid phase behavior of amino acids in solution that resulted in asymmetric amplification of one enantiomer. The reaction operates in aqueous media that can explain the high ee in biomolecules from a presumably racemic prebiotic world. Basic concepts of thermodynamically controlled solid-liquid phase behavior explain the amplification in ee. It seems that different ways exist to achieve enrichment of the enantiomers of amino acids: sublimation, evaporation, and mechanical separation (Blackmond et al. 2007). Sublimation appears to be a gas–solid transition and is usually a nonequilibrium process that might be of astrophysical relevance. Preferential sublimation of L-crystals of solid serine mixtures is the cause of the observed enantiomer enrichment. Striking differences in intermolecular interactions in enantiomers and racemates account for mechanisms capable of producing spontaneous separation of enantiomers.

A series of L-enriched amino acids (leucine, alanine, phenylalanine, valine, methionine, and serine) gave sublimates of higher ee upon heating in sublimation apparatus under vacuum (Cintas 2007). Hence, sublimation is a plausible mechanism for the formation of optically active crystals. The crystal state provides a stable means of maintaining the chiral integrity of chiral molecules. Gas–solid and solution-solid phase transitions both provide for a fractioning of enantiomer composition that may lead to highly enantiomer-enriched sublimates or solutions for some amino acids. The potential for evolution to a single solid-phase chiral state exists for enantiomers that form separate D- and L-crystals under ambient conditions. This is illustrated by aspartic acid solid crystals that are in equilibrium with solution of enantiomers under racemizing conditions. With enhanced attrition due to grinding, the characteristic sigmoidal shape of an autoinductive process was observed (Viedma et al. 2008).

Solid phase enrichment of enantiomers was faster in the presence of glass beads than in their absence at given temperatures (99% ee compared to 58% after 6 days). The data show that the emergence of a state of solid-phase single chirality for aspartic acid may be achieved in near equilibrium process even in the absence of mechanical stirring.

3.6 Asymmetric Organocatalysis: Special Importance of Chiral Proline

Routes by asymmetric organocatalysis are available to account for the occurrence of chiral molecules on earth. L-proline and L-serine yielded relatively high values of ee in aldol condensations (Cordova et al. 2005). An epimer of a distinct amino acid, for instance, proline or serine, can serve as a catalyst in aldol reactions (Cordova et al. 2005). *The percentage of the catalyzing amino acid determines the ee of the reaction product.* Thus, the preferred formation of D-ribose compared to that of L-ribose can be accounted for by the intrinsic property of the reaction system with an asymmetric molecule acting as biocatalyst. L-proline has exceptional properties due to its structure, Fig. 3.3.

Fig. 3.3 The structure of L-proline is shown. The *blue* line indicates the stereochemical feature of the chiral C-atom

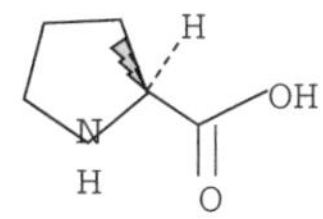

Proline is a stable, nontoxic, cyclic, secondary pyrrolidine-based amino acid with an increased pK_a value. Thus, proline is a chiral bidentate compound that can form catalytically active metal complexes (Melchiorre et al. 2008). Bidentate means that proline has not only one tooth but also a second one to bite and react. The greatest difference to other amino acids is a Lewis-base type catalysis that facilitates iminium and enamine-based reactions. It is especially noteworthy that cross-aldol condensations of unprotected glycoladehyde and racemic glyceraldehyde in the presence of catalytic amounts of the Zn-(proline)$_2$ gave a mixture of pentoses and hexoses (Kofoed et al. 2004). Again, proline seems to play the decisive role. The conditions are prebiotic: the reaction proceeded in water for seven days at room temperature. It is remarkable that the pentose products contained ribose (34%), lyxose (32%), arabinose (21%), and xylose (12%) and that all are stable under the conditions. Thus, the diastereomeric and enantiomeric selection observed support the idea that amino acids have been the source of chirality for prebiotic sugar synthesis.

3.7 Amplification of Ribonucleosides

A description of an amplification procedure based on the different solubility of the D-enantiomer and that of the corresponding D, L-racemate of ribonucleosides follows. When the melting points and solubilities of crystalline D, L-ribonucleosides and the pure D-ribonucleoside were determined, it was found that solution-based amplification of a slight ee of D-cytidine and D-adenosine, in a mixture with the D, L-racemates, is sufficiently large to produce solutions with at least 99% ee of the D-enantiomer (Breslow et al. 2010 and references cited therein). The 96% excess of D-uridine could also be sufficient to allow for the selection of the D isomer in solution under prebiotic conditions. In contrast, D, L-ribose itself forms a solid solution and D, L-guanosine a conglomerate (Breslow et al. 2010). This work is based on the mechanism for the amplification of fluctuations in racemic mixtures of the corresponding compounds (Morowitz 1969).

In this context, it is fascinating that 18 of the natural amino acids crystallize as racemates in a ratio of 1:1, except for threonine and arginine that crystallize as pure D- and pure L-crystals. Serine is particularly unusual as it is formed with an ee of 99% starting from an initial racemate in aqueous solution (Klussmann et al. 2006). Putting the data into perspective, one may conclude that it was likely that meteoritic impact delivered defined molecules from space to Earth. Among them, stereoisomers of α-methyl amino acids were in ee. These optically active compounds were the starting material for the transfer of chirality to other biomolecules. This

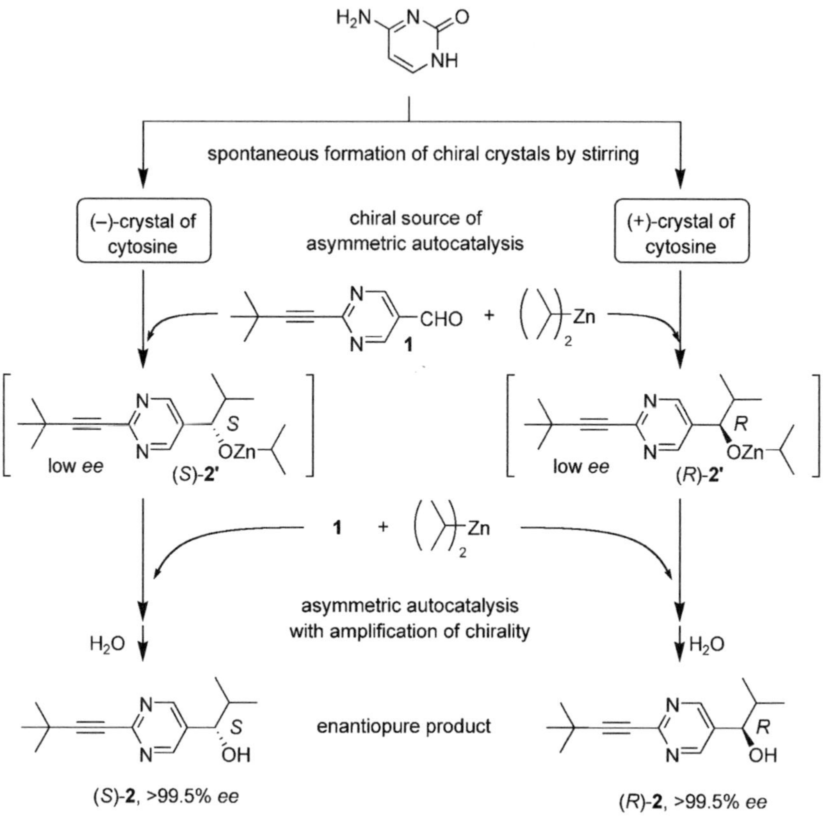

Fig. 3.4 The Soai reaction is shown that proceeds stepwise. The first step involves the reaction of Zinc-diisopropyl with a substituted aromatic aldehyde that yields a Zinc intermediate with low ee. The reaction product is autocatalytically amplified and has a chiral C-atom at the isopropyl group

could hold true for the constituents of proteins and for the components of lipids. This latter idea assumes that glycerol molecules delivered to Earth by meteorites were converted into the two stereoisomers glycerol-1-phosphate and glycerol-3-phosphate by corresponding ribozyme activities. However, the final experimental proof is still lacking. It is intriguing that in related systems, enantioselective phosphorylations of a carbohydrate derivate yielded enantiopure D-myo-1-inostol-phosphate by oligopeptides (Sculimbrene et al. 2002) compare with Fig. 3.4.

3.8 The Soai Reaction

In chemical synthesis, using chiral catalysts achieve a significant ee of either the left- or right-handed enantiomer. In these reactions, the ee of the chiral product is linearly related to the ee of the catalyst used. Some asymmetric reactions showed nonlinear behavior. Experimental evidence shows that autocatalytic processes can result in kinetically controlled asymmetric amplification. What is crucial in the reaction of chiral amplification is that dimers of the O-Zinc-diallkyl intermediate are the active catalysts. Racemic pyrimidine alcohols subjected to photolysis with either right or left-handed CPL produce an ee of one isomer. Direct asymmetric autocatalysis amplifies the slight excess of one enantiomer leading to enantiopure compound by reaction with diisopropylzinc (Soai et al. 1995). Soai's discovery of asymmetric amplification in the autocatalytic alkylation of pyrimidine aldehydes with dialkylzinc is remarkable. In this model reaction, addition of very low amounts of an ee of a catalyst of its alcohol products accelerates its reaction rate; this yields very highly enantioenriched catalyst as product as shown in the website: http://www.wikipedia.org.wiki/Homochirality.

This result supports the view that diverse ways exist to obtain chiral biomolecules via CPL or chiral inorganic or organic crystals combined with asymmetric autoctalysis. Kenso Soai and his team studied the effect of the structure of the substituents at position 2 of the pyrimidyl alkanol (Shibata et al. 1996). They found that using 2-alkynyl-pyrimidyl alkanol after three rounds of asymmetric autocatalysis, an astonishing amplification factor of 630,0000 was reached. In the reaction, either (+) or (−) crystals of Cytosine serve as initiators that were formed spontaneously by stirring. In the Soai reaction of chiral amplification, it is crucial that dimers of the O-Zinc diisopropyl intermediate are the active catalysts: Racemic pyrimidine alcohols subjected to photolysis with either right- or left-handed CPL produced an ee of one isomer as shown in Fig. 3.4.

Direct asymmetric autocatalysis amplified the slight excess of one enantiomer, leading to the enantiopure compound by reaction with diisopropylzinc. It is widely accepted that enantiomerically enriched products must form from achiral precursors merely because of statistical fluctuations. Usually, however, enantiomeric enrichment by fluctuations is very low. Thus, an amplification process of enantiomeric enrichment is required. Detailed kinetic analysis revealed that autocatalysis and inhibition are the major players in asymmetric autocatalytic synthesis. It turned out that tetramers serve as catalyst in the Soai reaction. The transition state for the Soai reaction implicates two molecules of pyrimidine alcohols or alcoxides as the dimeric catalysts and one molecule of prochiral aldehyde substrate (Buono and Blackmond 2003). Further kinetic studies using different ratios of substrate and reagent showed that a tetramer template is used.

Thus, *the Soai reaction is a template-directed self-replicating system that successfully maintains exponential growth kinetics and high autocatalytic efficiency over many turnovers*. The results support the view that multiple and diverse ways exist to obtain chiral biomolecules via CPL or chiral inorganic crystals such as quartz combined with asymmetric autoctalysis. It is, however, important to remember that the Soai reaction must be carried out in nonaqueous solvents under prebiotically unrealistic conditions.

3.9 Spontaneous Formation of Chiral Crystals of Cytosine During Mechanical Stirring

Cytosine is an essentially flat molecule. The three-dimensional structure of cytosine crystals revealed helices. It is involved in the Genetic Code of 17 amino acids and controls essential features of living systems. Cytosine can form under prebiotic conditions. Nonchiral cytosine spontaneously forms highly enantioenriched crystals upon stirring during crystallization. Furthermore, chiral crystals of cytosine act as chiral initiator for asymmetric autocatalysis with amplification of chirality to provide for a virtually enantiopure compound (Fig. 3.5).

Nonchiral cytosine when crystallized from methanol with stirring and without any seed yields powder-like crystals that either show a plus or minus Cotton-effect in the circular dichroism spectra (Kawasaki et al. 2008) Out of 55 stirred crystallizations, (+) cytosine formed 24 times, (−) cytosine 21 times, and ten samples were below the detection level (Fig. 3.5).

Based on the well-known 3-dimensional structure of cytosine crystals, a scenario of the evolution of chirality initiated by nonchiral cytosine seems likely. The X-ray structure of cytosine revealed the helical arrangement of the cytosine oligomers linked to each other by distinct hydrogen bonds. The chiral crystals can generate a large amount of enantiopure cytosine (>99.5% ee) when submitted to asymmetric autocatalysis. The reaction occurred by using the chiral crystal of cytosine as chiral initiator (Kawasaki et al. 2008).

Thus, the origin of homochirality of biomolecules might have involved the inherently achiral nucleotide base cytosine. In conjunction with the subsequent amplification of chirality by asymmetric autocatalysis, spontaneously formed chiral crystals of achiral cytosine acted as an origin of homochirality in biomolecules. The structure of cytosine indicates that the N1-atom may be prochiral (marked by an asterisk in Fig. 3.5). In contrast, uracil and adenosine do not have asymmetric atoms. It remains unknown whether the distantly related structure of guanine can

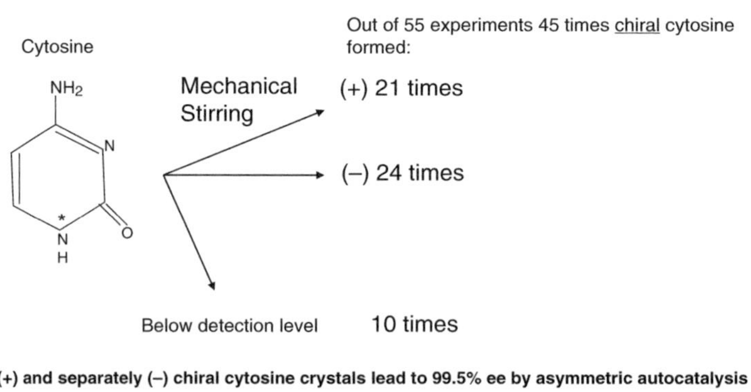

Fig. 3.5 Formation of chiral and achiral cytosine crystals is shown that occurs upon mechanical stirring of a solution in methanol

form chiral crystals by mechanical stirring. This may have technical difficulties due to the insolubility of guanine in water, and the much lower solubility in alcohols compared with cytosine. In any case, cytosine is one of the four building blocks of RNA. It is likely that the chiral form contributed to and predetermined the pathway to the RNA world, a scenario known to be of ancient origin as described and discussed by Gerald Joyce (2002).

3.10 The Butlerov or Formose Reaction

Butlerov found out that in alkaline medium (calcium hydroxide), formaldehyde HCHO polymerizes to form about 20 different sugars as racemic mixtures, Butlerov 1861. The reaction requires a divalent metal ion. Breslow found a detailed mechanism of reaction that explains the reaction products, (Breslow 1959). He found that glycol-aldehyde is the first product that is subsequently converted into glyceraldehyde (a triose), di-hydroxy-acetone, and then into various other sugars, tetrose, pentose, and hexose. *The formose reaction advances in an autocatalytic way in which the reaction product is itself the catalyst for that reaction with a long induction period.* The intermediary steps proceed via aldol and retro-aldol condensations and, in addition, keto-enol tautomerizations. It remains unexplained how the phosphorylation of 3-glyceraldehyde leads to glycral-3-phosphate (Fig. 3.6). Future work should study whether or not ribozymes exist that can carry out this reaction in a stereo-specific way.

When L-amino acids (with the exception of proline) catalyzed the formose reaction, an excess of D-glyceraldehyde formed. In contrast, without any special catalyst, an equal mixture of the D- and L-glyceraldehyde formed. The reaction conditions were prebiotic. Furthermore, addition of small amounts of water increased the enantiomeric excess to more than 90% (Breslow et al. 2010). An intriguing mechanism of chiral induction takes place when chiral L-amino acids (such as the ones found in the Murchison meteorite) were used as basic catalysts in the formose reaction induced about 10% ee of D-threose (Pizzarello and Weber 2004). Stereo-selective syntheses of pentose sugars occur under realistic prebiotic conditions when LL-dipeptides catalyzed the formose reaction (Pizzarello and Weber 2010).

According to Eschenmoser and his group, glycol-aldehyde- phosphate reacted with glyceraldehyde-2 phosphate to form ribose-2.4-di-phosphate as major product (Eschenmoser and Loewenthal 1992). The product also formed from formaldehyde and glycol-aldehyde-phosphate. Importantly, Eschenmoser's group obtained an ee of the ribose-phosphates. In this context, it is relevant that ribose formed upon reflux of formaldehyde over kaolinite. Ribose also formed from formaldehyde by brush discharges. D-ribose as starting material reacted with ortho-phosphate to form α-D-ribofuranosyl-1′-phoshate in a yield of about 15% in the presence of cyanogen $(CN)_2$ as condensation agent at 250°C (Ferris and Hagan 1984).

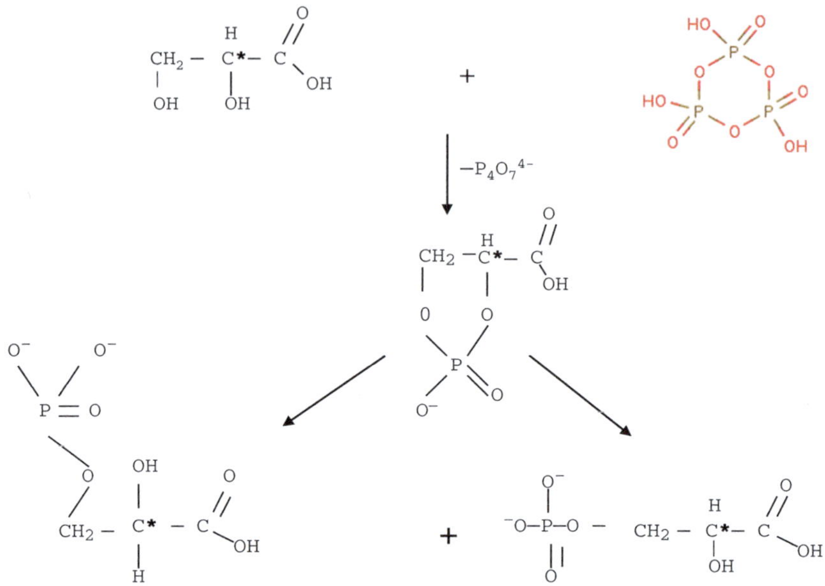

2-phospho-glyceraldehyde

3-phospho-glyceraldehyde

Fig. 3.6 The reaction of glyceraldehyde with trimetaphosphate (colored in *upper right corner*) is shown that results in the formation of the two isomers of phosphoglyceraldehyde. The chiral C-atom is marked with an *asterisk*

3.11 Conclusions from the Aspects of Chirality

Recent reports reveal several different ways to chiral amino acids, carbohydrates, and the chiral nucleobase cytosine. Intrinsic properties of the biomolecules such as the differences in solubilities, melting points, and vapor pressures of racemates and the corresponding pure enantiomers alone and combined gave rise to chirality on the ancient Earth. The data show how the start of chirality and life might have originated in diverse ways. In addition, asymmetric autocatalysis, mechanical stirring, and evaporation can lead to high ee of essential biomolecules under appropriate conditions. The chiral amino acids formed by evaporation from aqueous solutions or by sublimation corresponding to drying lagoons on the primordial Earth.

The formose reaction formed chiral sugar molecules when catalyzed by an appropriate chiral amino acid under credible prebiotic conditions. Alternatively, chiral amino acids likely formed via import from meteorites with subsequent transfer of chirality to other molecules. The solid chiral nucleobase cytosine arose by mechanical stirring from solution, Evaporation of ribonucleosides and subsequent hydrolysis led to chiral ribonucleosides.

The intrinsic differences of the physicochemical properties (for instance the solubilities) of racemates and pure enantiomers of biomolecules are responsible for the multiple origins of biomolecules. This aspect requires further intense research into the individual pathways. The origin of the chiral and biologically important glycerol-3-phosphate molecule remains unknown.

Chapter 4
RNAs and Their Constituents

Nucleic acids consist of three classes of compounds linked to each other in a specific way: one of four nucleobases, a carbohydrate, and phosphodiester. The basic building block consists of one of four nucleobases covalently linked to a ribofuranose sugar residue linked in turn via a bridge of a 3′-5′-phosphodiester to the next carbohydrate, see Chap. 8.

4.1 Nucleobases

Four nucleobases occur in RNA: adenine "A," uracil "U," guanine "G," and cytosine "C." Thymine, "T" replaces uracil in DNA. The synthesis of adenine discovered by Juan Oro is straightforward. HCN reacts with ammonia to yield ammonium cyanide. Upon heating in water, this compound forms adenine in high yield. Guanine, a closely related purine nucleobase, forms, although, in poor yield. There are several routes to the nucleobase cytosine. It forms by reaction of concentrated urea solutions with cyanacetaldehyde. Upon addition of water to cytosine, another nucleobase, uracil and ammonia form as shown in Fig. 2.1. Urea in turn was obtained from ammonium cyanate by heating, as discovered by Wöhler in 1928.

4.2 Carbohydrates

D-ribose, a 5-membered cyclic furanose is the characteristic sugar building block of RNA. Its main features are two adjacent or vicinal OH-groups located at the carbon atoms in position 2 and 3. This particular feature explains why RNA is labile toward alkaline agents. Borate and Mg^{2+} ions stabilize the vicinal OH groups by forming complexes with them. The other characteristic difference of RNA compared to DNA is that an H atom replaces the OH-group at position 2 of the sugar in DNA. In 1861, Butlerov found out that in alkaline medium (calcium hydroxide), formaldehyde HCHO polymerizes to form about 20 different sugars as racemic mixtures, see also Sect. 3.10. The formose reaction requires a divalent metal ion, and when

R.M. Flügel, *Chirality and Life*, DOI 10.1007/978-3-642-16977-9_4,

reproduced and analyzed in detail, it was found that glycolaldehyde is the first product that is subsequently converted into glyceraldehyde (a triose), dihydroxyacetone, and then into various other sugars, tetrose, pentose, and hexose. It advances in an autocatalytic way in which the reaction product is itself the catalyst for that reaction with a long induction period. The intermediary steps proceed via aldol and retro-aldol condensations and, in addition, keto-enol tautomerizations, see Sect. 3.10. When L-amino acids (with the exception of proline) catalyzed the formose reaction, an excess of D-glyceraldehyde formed. Without any special catalyst, an equal mixture of the D- and L-glyceraldehyde formed. The reaction conditions were prebiotic. Furthermore, addition of small amounts of water increased the enatiomericexcess to more than 90%. Glycol-aldehyde-phosphate reacted with glyceraldehyde-2 phosphate, to form ribose-2.4-di-phosphate as major product. This product also forms from formaldehyde and glycolaldehyde-phosphate. Importantly, Eschenmoser's group obtained an enantiomeric excess of the ribose-phosphates. In this context, it is relevant that ribose formed upon reflux of formaldehyde over kaolinite. Ribose also formed from formaldehyde by brush discharges. D-ribose as starting material reacted with orthophosphate to form α-D-ribofurnosyl-1′-phoshate in a yield of about 15% in the presence of cyanogen $(CN)_2$ as condensation agent at 250°C. It remained, however, an enigma how natural ribonucleoside-5′-phosphates were generated under the conditions of the prebiotic soup. Only recently, a British group succeeded in synthesizing activated pyrimidine ribonucleotides. The staring materials were cyan amide, cyan-acetylene, glycolaldehyde, glyceraldehyde, and inorganic phosphate. Sutherland's group showed that a single 2-amino-oxazole intermediate contributes atoms to both the sugar and the nucleobase (Powner et al. 2009).

4.3 Phosphodiester

Nature designed many different forms of esters in which phosphorous acid links to diverse groups of OH-groups of alcohols, sugars, and other compounds. 3′-5′-phosphodiesters are the predominant linkages in the both RNA and DNA. Phosphodiesters also occur in membranes and coenzymes. These routes of synthesis indirectly open up the path to nucleotides. Did compounds of phosphorus play an essential role in the very first phases of the beginning of living systems? This question has to remain open. For instance, Kolb and Orgel (1996) found that glyceric acid reacts with cyclic trimetaphosphate to yield the racemic mixture of 2- and 3-phosphoglycerol acid in good yields as schematically shown in Fig. 3.6. Note that the racemates form in the reactions. The asterisks indicate the chiral nature of the carbon atoms. As recently proposed, the geochemistry of phosphorous on the early Earth was controlled by reduced oxidation states of P compounds, such as the doubly negatively charged phosphite ion HPO_3^{2-}. The doubly negatively charged phosphite is more soluble and reactive than orthophosphates. The reduced oxidation state of P originated from extraterrestrial material that fell during the

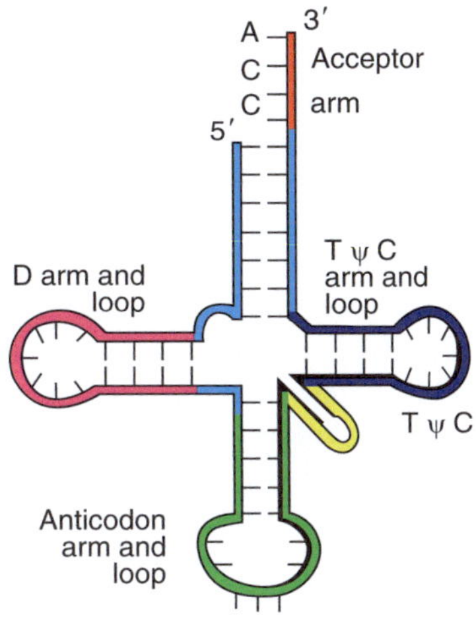

Fig. 4.1 The cloverleaf structure of transfer RNA is shown. The main features of the arms are colored: the anticodon arm in *green* that pairs with messenger RNA during translation, The acceptor stem (*red*) with the protruding CCA-end to which the activated amino acids are attached

heavy bombardment period and persisted in the mildly reducing atmosphere. There is experimental evidence for this view, providing an alternate route to the formation of prebiotic P compounds. The mineral schreibersite (Fe, Ni)$_3$P, which was also shown to occur in meteorites, significantly influenced the geochemistry of P, which led to the reaction of acetate with this mineral to form C–P-containing compounds, such as acetylphosphonate, hydroxymethylphosphonate, and phosphoglycolate.

4.4 Classes of RNA

4.4.1 Transfer RNA

Transfer (tRNA) is the information adapter molecule. It is the direct interface between the amino-acid sequence of a protein and the information in DNA. Therefore, it decodes the information in DNA. The different tRNA molecules have between 75 and 95 nucleotides. Transfer RNAs from all organisms have a similar structure; indeed, a human tRNA can function in yeast cells.

Figure 4.1 depicts the cloverleaf structure of a tRNA; the bars represent base pairs in the stems. There are four arms and three loops – the acceptor, D, T pseudouridine C, and anticodon arms, and D, T pseudouridine C, and anticodon loops. Sometimes tRNA molecules have an extra or variable loop (shown in yellow in Fig. 4.1). The synthesis of transfer RNA proceeds in two steps. The body of the tRNA is transcribed from a tRNA gene. The acceptor stem is the same for all tRNA molecules and added after the synthesis of the main body. It is replaced often during lifetime of a tRNA molecule. The 3-D structure of a yeast tRNA molecule, which can code for the amino acid serine, shows how the molecule is folded with the

D and T pseudo-U C loops in contact and with the acceptor stem and the anticodon loop at opposite ends. This folding results in the familiar "L" shape of tRNA. The L shape is detectable in crystallographic analyses by X-ray diffraction. The acceptor stem is the site at which a specific amino acid is attached by an enzyme, an amino-acyl-tRNA synthase. The anticodon reads the information in an mRNA sequence by base pairing. A commonly observed motif is the U-turn, seen in the anticodon loop of tRNA, which involves hydrogen bonding of the N3 position of a uridine with the phosphate group of a nucleotide three positions downstream, causing an abrupt reversal in the direction of the tRNA chain. The activation of the 2′- or 3′-OH by AMP-amino acid intermediates is the crucial step in the emergence of translation and the evolution of a genetic code (this will be discussed below).

A completely different function of a tRNA is to serve as primer for reverse transcription of retroviral RNA into DNA. This might well be a hint for a relic from the RNA world. Estimates for the age of the first tRNA molecules by Eigen and his group indicate 3.8 billion years. It would take too long to enumerate all the known functions of tRNA, for instance, regulation of splicing in *trans*-position.

4.4.2 Ribosomal RNA

The resolution of the 3-D structure of the small and large subunit of ribosomal RNA (rRNA) revealed an intricate, globular secondary structure with extensive base-stacking interactions. Forming stems and loops show that rRNA does not serve as a scaffold alone; however, in fact, it also creates the framework for the ribosome and the functional sites to act as a true natural ribozyme. The functional sites are located in the center of the ribozyme. It has been said that tRNA looks like Nature's attempt to make RNA do the job of a protein; rRNA takes this notion one giant step further to make a functional molecule out of RNA. Beyond this, Nature has to resort to proteins.

Within the ribosome, RNA helices are irregularly packed against one another to form a globular overall structure. It is striking that ribosomal proteins are absent near the functional site of the peptidyl transferase centre.

4.4.3 Messenger RNA

A process called transcription generates messenger RNAs (mRNA). A large enzyme, RNA polymerase II, functions as a catalyst to synthesize long chains of single-stranded ribonucleic acids in the presence of Mg^{2+} ions and the four rNTPs. It codes for proteins and occurs in almost all sizes.

Out of one pre-mRNA molecule, the process of splicing forms differently spliced versions of shorter RNAs. By translation, the spliced RNAs lead to various proteins.

Most prokaryotic transcripts originate using adenosine-5′-triphosphate and, to a lesser extent, guanosine-5′-triphosphate at the start site of the growing RNA chain.

Thus, the 5′-ends of RNA chains start with the characteristic triphosphate residue in contrast to the DNA polymerases that require an additional RNA primer molecule for initiation. The 5′-end phosphate groups protect the RNAs against degradation against ubiquitously occurring ribonucleases. Eukaryotic transcripts carry a special structure called cap at their 5′-end. The 5′-cap contains a modified nucleotide at the 5′-end of precursor messenger RNA in a 5′-5′-phosphodiester linkage. The process of 5′-capping is vital to creating mature messenger RNAs that translation converts into proteins. The 5′-caps protect the RNAs against degradation by ribonucleases, too. Poly (A)-addition enzymes modify at their 3′-termini. The poly (A)-ends protect mRNAs against degradation and furthermore serve as signal for cellular transport from the cell nucleus to the cytoplasm.

4.4.4 Micro and Small RNAs

Many classes of small, double-stranded, noncoding RNA molecules serve to regulate gene expression. Among them are the short noncoding micro-RNAs, silencing RNAs, small nuclear and small nucleolar RNAs, and some more. The small nuclear RNAs are an essential component of the spliceosome and play a decisive during splicing. The spliceosome itself carries out a complex set of reactions that apparently require many proteins. However, some steps proceed with RNAs in the absence of protein. Thus, it has to be a ribozyme.

4.4.5 RNA Modifications

Besides capping and splicing, an additional process changes the readout of any mRNA. By a process called RNA editing, bases at the 3′-end can be inserted and deleted. The diversity of RNA editing mechanisms includes nucleoside modifications such as C to U and A to Inosine deaminations, as well as nontemplated nucleotide additions and insertions. RNA editing in mRNAs effectively alters the amino acid sequence of the encoded protein so that it differs from that predicted by the genomic DNA sequence. It is unknown whether most of the RNA editing processes, which also include tRNA and rRNA, are evolutionarily recent acquisitions that arose independently.

A small percentage of adenosine residues in any mRNA carry a methyl group at the 6-amino group of the nucleobase. About 200 different residues occur in most transfer RNAs where they contribute to its secondary structure. The website at http://library.med.utah.edu/RNAmods/imaover.htm is useful. More RNA modifications include small molecule – RNA conjugates. One is coenzyme A added to small RNAs.

Chapter 5
Proteins and Their Constituents

5.1 Amino Acids

Amino acids are the basic building blocks of peptides and proteins and are simpler in structure than ribonucleotides. They are the major constituents for building up cells and serve as energy storage, too. Peptides and proteins mainly occur as L-enantiomers in nature. Almost each enantiomer of any amino acid can act as a specific enzyme in aldol condensations.

The D-enantiomeric amino acids form by nonprotein synthesis inside bacteria in the form of peptide antibiotics. Although proteins predominate in today's DNA world, one assumes that proteins were in a minority at the beginning of the RNA world. It is likely that several pathways independently produced the first smaller peptides early on. Many essential coenzymes consist not only of nucleotide derivatives but also amino acids: Coenzyme A, $NADH^+$, S-adenosyl-methionine, tetrahydrofolate, and many more, see below.

The following list summarizes those properties that distinguish proteins from RNA.

1. Proteins consist of 20, in some bacteria 22, different amino acids covalently linked with each other by peptide bonds. The amino acid side chains form many hydrophobic bonds, intermolecular hydrogen bridge bonds, and salt bridges within the protein chains or with another partner molecule.
2. The main types of the dynamic structures of protein encompass three classes: the secondary-, tertiary-, and quarternary structure.
   Stable protein structures require amino acids all of one configuration only, namely the L-configuration but not a mixture of L- and D-enantiomers.
3. They are relatively stable against heat, UV-irradiation, acid, and basic hydrolysis.
4. Proteins have an unusually large, seemingly unlimited repertoire of enzymatic and other activities.
5. They are very versatile in receiving and sending internal and external signals including different partner cells, in particular in association with biomembranes. It would take too large a space to list all the diverse functional properties of the proteins here.

R.M. Flügel, *Chirality and Life*, DOI 10.1007/978-3-642-16977-9_5,

6. A special class of proteins, prions, can form stable and transmissible self-aggregates.
7. Proteins consist mainly of C, H, N, O, S, Se, and can bind diverse metal ions in modified side-chains that often carry phosphate residues. These modified residues often lead to changes in structure and function.
8. Proteins occur in all sizes. The amino terminus may often carry an acyl-group, whereas the COOH-terminus may carry an amide group.
9. Amino acids have a much more diverse set of functional groups than nucleotides. The functional groups of the side chains of the amino acids include carboxylate, carboxamide, imide, hydroxy, thiol, and primary and secondary amines. These functional groups are responsible for the numerous interactions in peptides and proteins. When they are part of large proteins, they display three-dimensional structures. The repertoire may be even larger by forming special active sites.

Somc claim that there is no doubt that proteins that formed easily were first on the scene. Of course, the first proteins must have been much shorter than any used in life today, because of the sheer unlikelihood of forming useful long ones out of a mixture of amino acids.

It is unlikely that the natural amino acids formed in one set of reactions as the structure of distinct amino acids form a range of structures. Glycine is achiral and the least complex. Phenylalanine, tyrosine, tryptophane, selenocystine, and pyrrololysine are the most complex members. Some amino acids are derivatives of individual ones. For instance, asparagine is the amide of aspartic acid, glutamine the amide of glutamic acid. One assumes that several factors guided the formation of distinct amino acids. Prebiotic synthesis was an inadequate source of the 22 amino acids. Instead, some of them formed from coevolving pathways of amino acid biosynthesis.

Meteoritic impacts delivered natural and "unnatural" amino acids to Earth, especially in the early phases. The prebiotic soup and/or the FeS scenario produce amino acids as racemates. In contrast to the RNA world, diverse reactions in five different scenarios generated peptides.

Importantly, salt-induced peptide formation could provide an abiotic route for the formation of peptides directly from amino acids in concentrated NaCl solutions containing copper ions. Montmorillonite and similar minerals apparently promote the condensation reaction that could have taken place in evaporating tidal pools – Darwin's warm little ponds – where the required salty brine solutions were easily available. Obviously, this is a likely and hence a credible prebiotic scenario. There might a pearl hidden beneath muddy waters. Besides, it is fascinating to assume that the primitive enolase enzyme known to be a highly conserved ancient enzyme could have evolved in an RNA-peptide world. Enolase catalyzes the for enantio-selective carbon–carbon bond addition of water to phosphoenol pyruvate to yield D-2-phospho-glycerate.

Now back to the different routes of prebiotic peptide synthesis.

(a) Meteorites provided for the extraterrestrial delivery of amino acids. The Strecker synthesis is responsible for forming amino acids from the corresponding aldehydes and ammonia. Experiments carried out under conditions of deep space produce many amino acids with simple compounds as interstellar ice starting material: Using interstellar ice materials H_2O, methanol CH_3OH, NH_3, CO, and CO_2 in a molar ratio of 2:1:1:1:1 and UV-irradiation at 12 K in a high vacuum. After acid hydrolysis at room temperature, the main products were racemic mixtures of Gly, Ala, Ser, Val, Pro, Asp, 15 nonstandard amino acids, and several other compounds.

(b) CO-driven fixation in a reaction catalyzed by iron-nickel sulfide under conditions of hydrothermal, volcanic conditions.

(c) Different synthetic routes produce amino acids under credible prebiotic conditions but they form racemic mixtures in all cases.

(d) American researchers succeeded in synthesizing long prebiotic oligomers of up to 55 amino acids on hydroxy-apatite surfaces.

(e) Quite recently, experiments that simulated the conditions that dominated when a meteorite hit the Earth's surface showed that a metal coin shot at high energy with a can-sized bullet that contained water and the 20 natural amino acids in an experiment performed at an angle of 25° corresponding to the angle of meteorite impacts. Subsequent analyses revealed the presence of di-, tri-, and tetra-peptides. Thus, meteorite impact generated small peptides. The last result is of special interest as the chirality of the peptide products reflects that of the input amino acid. This would lead to optically active peptides. This is not the case for peptide formation from amino acids in aqueous solution since they rely on condensation reactions; they are both thermodynamically and kinetically unfavorable.

There are different fashions how to induce water removal in condensation reactions. By simply heating amino acids with or without potent agents of condensations such as hydrogen cyanide, HC≡N, cyan amide, N≡C–NH_2, and carbon-oxy-sulfide COS, oligomers and polymers, called proteinoids, readily formed . The detection of enzymatic activities in these polymeric proteinoids was unsuccessful except for about ten degrading enzymatic activities. Synthetic activities that build up molecules, for instance, kinase, ligase, and polymerase, were not detectable. The disadvantage of these mixtures of proteinoids is that they do not exhibit a distinct structure or a function.

Significantly, they lack chirality except for when one starts with natural amino acids. Now once peptides or poly-(amino) acids formed by one of the other reactions it was shown that they could have served as catalysts for the amplification in asymmetric reactions. As shown by numerous experiments, clays, metal cations, the base imidazole, and HCN derivatives might have served as efficient condensation agents in catalyzed oligomerization and polymerizations. Even simple dipeptides can catalyze the addition of HCN to aldehydes to give optically active cyanohydrins. This might be one way to an asymmetric system with considerable prebiotic relevance. Simple peptides with a sequence of about 20 amino acids can adopt a unique protein secondary structure since upon tetramerization (that is the folding into structure of four identical subunits), it rapidly folded into four-helix bundle with a tight core. Furthermore, asymmetric acylation with tetrapeptides were successfully performed, and on top of all, the X-ray structure of a phenylalanine-proline-isoamino-buturyic acid derivative revealed a type II-beta turns.

http://www.ncbi.nlm.nih.gov/

The NIH databank covers a wide range of information on sequences of genomes and proteins of many species. In addition, it offers a public library of scientific literature and many other items.

http://www.isb-sib.ch/

The famous Swiss databank is versatile and especially useful for finding properties and sequences of proteins.

http://www.ebi.ac.uk/embl/

The EMBL databank covers many genomic and protein sequences and additionally offers special services for searching phylogenetic relatedness between proteins.

5.2 Posttranslational Modification

There are virtually countless forms of posttranslational modifications so that the number of structural and implicitly, the number of functional information increases endlessly.

5.3 Protein Structural Motifs: α-Helix, β-Sheet, β-Turns

The *secondary structure* types are the α-helix, β-sheet, and β-turn. The *tertiary* structure groups are the overall shape of a single protein molecule; the spatial relationship of structures to one another. The term "tertiary structure" is often used as synonymous with the term fold, the shape or structure that results from the interaction of more than one protein molecule, usually called protein subunits in this context, which function as part of the larger assembly or protein complex. The *quarternary structure*: the shape or structure that results from the nteraction of more than one protein molecule, usually called protein subunits in this context, which function as part of the larger assembly or protein complex.

5.4 Protein Domains

Protein domains are conserved regions of a limited number of amino acids that can bind diverse partner molecules to form structures of a higher complexity. The domains have a heterogeneous internal organization that consist of amino acid interactions and comprise multiple functionally distinct sectors or subdivisions. The sectors are regions of proteins that are distinct from the hierarchy of primary, secondary, tertiary, and quaternary structures.

Chapter 6
Membranes

6.1 Long Chain Fatty Acids Esterified with Phosphoryl Groups: Lipid Bilayer of Bacterial Membranes

The structures of the esters are shown in the upper rows of Fig. 6.2.

6.2 Prenyl Chains Linked to Di-Isoprene Phosphate via Ether Bridges: Lipid Mono- and Bilayers of Archaeal Membranes

Precursors of membrane components, carbohydrates, amino acids, and other biogenic compounds formed either directly or indirectly from those chiral and nonchiral molecules delivered to Earth. Benzene, for instance, is one of the aromatic compounds of interstellar dust from which a precursor of the membrane lipid inositol-1-phosphate apparently derived by oxidation in a long series of chemical reactions. The structure of the important lipid, a derivative of inositol-*phosphate* in Fig. 6.1, displays two different fatty acid side chains (marked in blue and green) that frequently occur in membranes.

The numbering of the cyclohexanol ring is different from the regular numbering of the glycerol part to which long chain fatty acids link through a phosphate bridge. The four arrows indicate the chiral carbon atom No. 2 (marked by a red asterisk) as well as the relative spatial arrangements of the groups.

The formation of ether-type polar lipids that occur in the membranes of Archaea remains elusive. The enantiomeric glycerol-phosphate backbone, ether linkages, and isoprenoid chains are distinguishing features of archaeal lipids. Carbonaceous meteorites contain up to several percent of their mass as organic carbon, mainly polycyclic aromatic hydrocarbons. Material extracted from the Murchison meteorite by organic solvents contains amphiphiles that form membrane-like vesicles in aqueous solution.

The flux of energy into the organization of matter appears virtually or hardly possible without compartments. Thus, the formation of primitive membranes was the decisive step to develop protocells, since a membrane warrants a safe protection

R.M. Flügel, *Chirality and Life*, DOI 10.1007/978-3-642-16977-9_6,

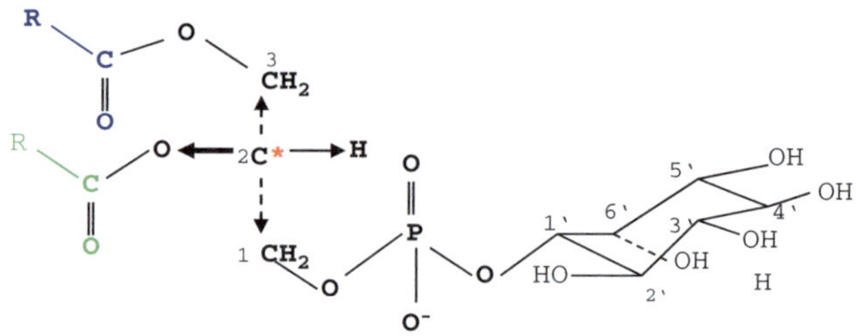

Fig. 6.1 The stereochemical features of inositol-phosphate are shown. The *right hand part* of the molecule presents the chair conformation of the hydroxy-cyclohexanol part attached to the 1′-OH group via a phosphate group to the glycerol part. The chiral C2-atom is marked with an *asterisk*. The C1- and the C3-atoms of glycerol are connected to two different long chain fatty acids (in *green* and *blue*)

of all sensitive substances against degradation within its confinement. Moreover, the spontaneous formation of a lipid double layer in water enables specific transport processes even in vesicle-like primitive and model membranes.

Specific transport of distinct ions and low molecular mass compounds is another prerequisite for a further development of protocells. Large molecules form within membranes as they cannot enter from the outside. The polar double layer lipid membrane is permeable for water and neutral lipids. In all events, biomembranes are crucial for formation and maintenance of primitive protocells.

The following properties of membranes are independent of the precise type of membrane. Membranes are permeable for water and neutral lipophilic molecules. They are much less permeable for polar compounds such as sugars, and even less permeable for inorganic ions. They have a high electric resistance. The structure of bacterial biomembranes that consist of a continuous double lipid layer makes it easy to understand its function as barrier. The two essential domains within a lipid are the outer hydrophilic head region and the hydrophobic or nonpolar tail region that is located at the inside (see Fig. 6.2). Water surrounds the hydrophilic head region, while the hydrophobic domain is repulsive against water. Thus, in the lipid bilayers, the hydrocarbon parts of the tails are in opposition to each other forming a hydrophobic core, while the positively and negatively charged heads project into the aqueous solution on both sides of the membrane. The lipid-containing core accounts for the permeability for small hydrophobic substances and the decreased permeability for hydrophilic molecules. In aqueous media, phospholipids tend to form lipid bilayers spontaneously. Therefore, the overall properties of a lipid bilayer membrane illustrate an innovative leap of quality. To maintain the bilayer configuration, an energy supply is not required. The interaction of simple, single-chain amphiphile molecules with many different surfaces of mineral particles results in the organization of membranes and in the formation of vesicles.

The structure, building up, and functions of diverse membranes of cells, organelles, and vesicles of all organisms concern purely physical–chemical issues, although they are quite complex and diverse. The essential structural components of membranes are the lipids; in most cases, phospholipids usually contain proteins

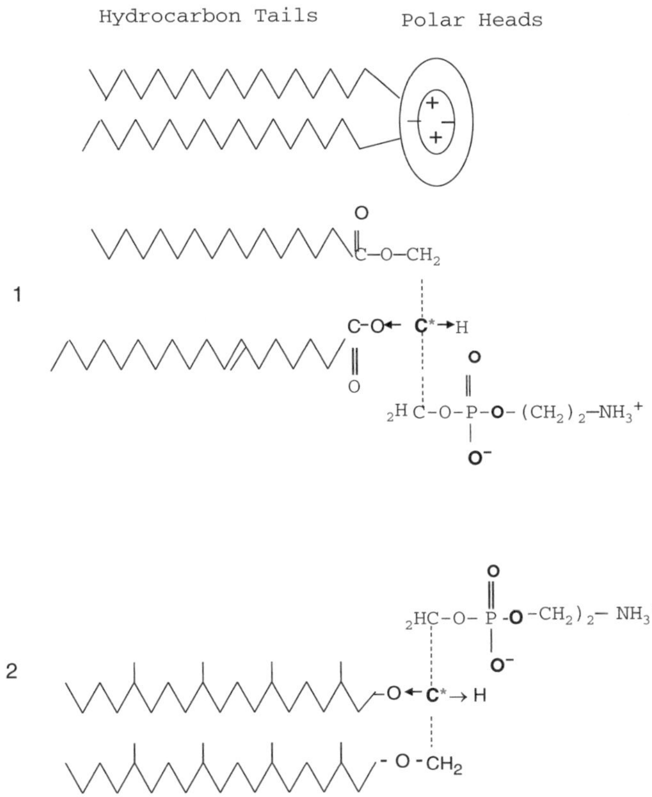

Fig. 6.2 The *upper part* shows schematically the typical feature of a lipid with a polar head and a longer hydrocarbon tail. In row 1, the glycerol with the chiral C-atom marked with an *asterisk* is esterified with a saturated fatty acid (*upper part*), and the *lower part* shows an ester with an unsaturated fatty acid. The polar head here consists of phosphate connected to an amino group. Row 2 presents an ether linkage as it occurs in Archaea

embedded within the lipid bilayer without forming covalent bonds between the lipid and protein parts. Most biomembranes contain about 40% lipids and 60% proteins. The basic building blocks of contemporary water-insoluble lipids are saturated and unsaturated fatty acids in *cis* configuration with a chain length of about 14–24, mostly 16–18 carbon atoms. Fatty acids with short tails, for instance, 9 C-atoms, do not form a bilayer. Thioester condensations of acetyl-CoA or its precursors served as the starting material for the synthesis of lipids.

Tri-acyl glycerides or neutral lipids contain three fatty acids esterified with the hydroxy groups of glycerol. The phosphoglycerides belong to the amphipathic and polar lipids that contain two fatty acid residues linked to the alcohol groups of glycerol and, in addition, the secondary alcohol group form ester with a phosphate residue.

The naturally occurring phosphoglycerides are optically active. Polar lipids spontaneously form micelles, mono- and bimolecular layers considered to represent a primitive kind of cellular membranes.

The structural unit of the phosphoglycerides is D-glycerol-3-phosphate, a derivative of the trivalent alcohol glycerol. The compounds glycerol and di-hydroxy-acetone were detectable in the Murchison meteorite. This does seem noteworthy as these biogenic compounds are not in the products of the prebiotic soup. What is the mechanism of lipid formation in meteorites? This question has not been answered.

There are diverse types of phospholipids. Notwithstanding their diverse structures, it is crucial that they all carry positive and negative charges at their ends.

Concerning archaeal lipids, we recognize four structural characteristics that are distinct from their counterparts in Bacteria and Eukaryotes. First, the most important difference is that the stereochemical structure of the glycerol-phosphate backbone of the Archaeal phospholipids is glycerol-1-phosphate. Glycerol-1-phosphate is the enantiomer of the of the glycerol-3-phosphate backbone of bacterial and eukaryotic phospholipids (Pereto et al. 2004).

The second feature is that the hydrocarbon chains link to the glycerol moiety exclusively by ether linkages in Archaeal lipids. This is in contrast to the Bacterial polar lipids most of which have ester linkages between the fatty acids and the glycerol residue. From the point of view of simplicity, it seems that mesophilic Bacteria existed before the Archaea since the latter contain more complex lipids.

Ethers are compounds in which an oxygen atom link to two substituted alkyl or aryl groups. The general formula for any ether is R–O–R′. Ethers are much more stable against hydrolysis than esters, an indication of why thermophilic Archaea contain ether linkages in their membranes. The third hallmark is that the hydrocarbon chains of polar lipids are isoprenoid derivatives containing regularly spaced methyl groups. The fourth difference is that in bipolar lipids of many Archaea, tetra-ether cores span the membrane to form a membrane monolayer.

Figure 6.2 schematically shows the structure of a lipid with one polar head and two long hydrophobic tails connected by a three-carbon backbone (uppermost row). Row 1 shows a diacyl lipid with two different carbonic esters. Row 2 illustrates an example for a lipid from Archaea with ether linkages.

The structure of inositol shown in Fig. 6.1 is another typical widely occurring membrane lipid.

It is unknown how the fatty chain carbonic acids of the lipids originally formed. It is noteworthy that fatty acids with up to six and seven C-atoms were detectable in the Murchison meteorite. Thus, it might be possible that membranes that were even more primitive existed in the past. However, dicarboxylic acids of chain lengths up to 17 were detectable after hydrothermal treatment. It is also remarkable that isoprenoid derivatives – more precisely, phytanyl and biphytanyl, that is C10 and C20 chains – linked to each other by monoether or diether bridges are part of membranes of the Archaea (Fig. 6.2, bottom structure).

It seems that Archaea entered their hot habitats at later phases, after mesophilic, double-membrane bacteria had developed. The term Archaea is misleading as genuine mesophilic Bacteria existed before Archaea. It is also counterintuitive that monolayer bacteria formed from the double-membrane bacteria by losing one membrane.

Another factor adds to an even more stunning complexity. This is due to the real structures of biomembranes. Built on the basic lipid bilayer, they insert a multitude of transmembrane proteins as well as diverse sugars with linear and branched structures into membrane that, in turn, can bind other proteins. Everybody who wishes to get a deeper insight into cell membranes should have a look at the Web site http://en.wikipedia.org7wiki/Cell_membrane.

Thus, a first living system developed within natural, preformed, and three-dimensional compartments in warm hydrothermal, submarine springs. If combined with the FeS world, it looks like an attractive idea. Actually, natural inorganic compartments exist in old springs of the oceans and gave high credibility to this idea. The compartments ranged in size from about 1.0 up to 100 μm. The main question remains how today's membranes formed from these pore-like inorganic mineral structures.

Prebiotic synthesis likely contributed to the spectrum of compounds in a relative early phase of the Earth since exposure of methane and water to brush discharges resulted in the formation of monocarboxylic acids up to a length of 12 carbon atoms. The fact that essential lipids do have optical activity might indicate that they are of meteoritic origin.

Distinct proteins noncovalently associated within biomembranes are responsible for the specific permeability of different cations and anions. They are termed channel molecules because of their pore-forming capability. New models of the lipid bilayer membrane assume that distinct proteins, such as specific receptors, penetrated into and are dynamically embedded in the lipid layer in more or less regular distances or they fill in its leaky gaps. Many proteins completely extend through the double layer and selectively allow the selective transport of distinct ions to the inside. Without much effort, these mainly hydrophobic interactions are consistent with the overall properties of natural biomembranes. Furthermore, the surface molecules serve to receive external signals, transmitting them to the interior of the cell and to communicate with other cells. The locations of these receptor and channel proteins within a given membrane are not fixed but dynamic. Extensive phylogenetic analyses of small but ancient phospholipid proteins, the ATPases, showed that they contain a hydrophobic alpha helical structure that would partition with the membrane phospholipid bilayer.

An approach to prepare vesicles consists of preparing a system of self-reproducing fatty acid vesicle in which enzymatic synthesis of polyadenylic acid takes place. In summary, a template-directed synthesis of a genetic polymer in a model protocell was successful. Vesicles are useful model systems of primitive forms of protomembranes. They possess important features of true membranes. The permeability for water and neutral lipophilic molecules and the virtual nonpermeability for polar compounds and inorganic ions are relatively similar. They simulate the structure of biomembranes displaying a virtually continuous double lipid layer. Thus, it appears that a combination of vesicles that contain diverse biopolymers by way of fusion provides a reasonable solution to the forming of forerunners of protocells. Membranes can carry out the processes required for living: they breathe, transport, store energy, receive, and send information.

Chapter 7
Compounds Required for Life

7.1 Water

Liquid water with its unique properties and universal solubility for many compounds indirectly required some fixed external parameters, for instance, the preferred range of temperature, geological times, and tolerated doses of UV-radiation. The amount of liquid water makes up the crucial difference between Earth and the other planets of our solar system. It also enabled the plate system to operate.

7.2 Coenzymes

Coenzymes have many functions. They carry chemical energy in their easily hydrolyzed acid-anhydride bond, for instance, ATP. ATP is one of the most important coenzymes as shown in Fig. 7.1.

The C1′ of D-ribose links N9 atom of adenine via a β N-glycosidic bond. ATP contains four chiral carbon atoms (C1′ to C4′) in its sugar moiety. Broken and thick arrows indicate the stereochemical features.

Figure 7.1 shows D-α-ribonucleotide-adenosine-5′-triphophate. The purine nucleobase adenine is in blue, D-ribose is in grey with two secondary, vicinal OH-groups are at the 2′ and 3′-positions, and three phosphate groups are in brown linked to the primary 5′-OH of the sugar.

Coenzyme A contains one chiral center at the secondary OH group of dihydroxy-dimethyl-butancarbonic acid. They combine with other groups to form enzymes; an example is coenzyme A. Figure 7.2 shows the structure of coenzyme A that consists of the nucleobase adenine, the sugar ribose linked via the C5′ OH group to a phosphodiester, dihydroxy-dimethyl-butancarbonic acid, and β-mercapto-ethylamine. It ends in a reactive thioester with an energy-rich bond.

Acetyl-CoA is necessary for the synthesis of amino acids, carbohydrates, nucleotides, and lipids. Thus, acetyl-CoA, the primary but not primordial CO_2 acceptor molecule, serves as a paradigm for almost every aspect of the molecular secrets of ancient pathways.

R.M. Flügel, *Chirality and Life*, DOI 10.1007/978-3-642-16977-9_7,

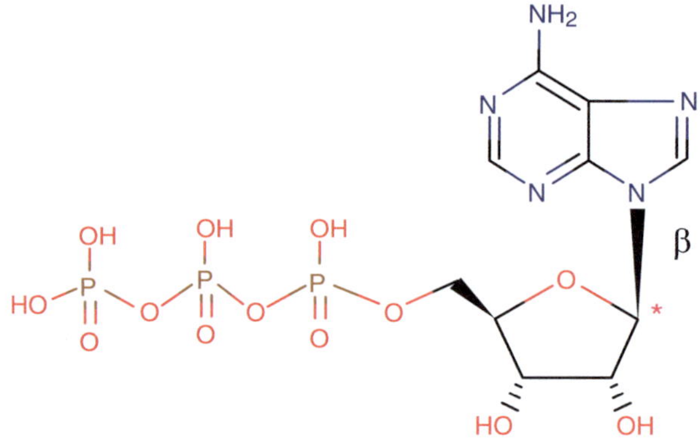

Fig. 7.1 The structural formula of ATP is shown with its main characteristics. The purine part is in *blue* that is connected in a beta glycosidic bond to the furanose sugar moiety; the OH groups are in *red* as are the three phosphate groups that are attached to the 5'OH group of the ribose

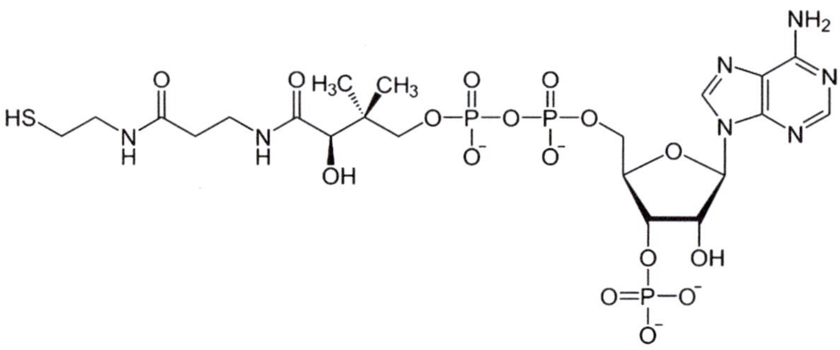

Fig. 7.2 The complex structure of coenzyme A displays a phosporylated adenosine part. The diphosphate bridge connects to a substituted ethanolamine derivative that ends in a reactive SH-group. The broken and thick lines indicate the special stereochemical feature of CoA

Coenzymes are used as specific signaling molecules in a cell, for instance, cyclic AMP. Energy sources of high and low level were present at the ancient Earth. They encompass sunlight, UV-radiation, volcanos, hydrothermal geysers and vents, electric discharges, energy from chemical reactions, and radioactive decay. Some biogenic compounds with energy-rich bonds such as ATP and thioesters contributed to chemical reactions. A description of the major classes of biopolymers and their constituents follows. The prebiotic synthesis of the building blocks of the major classes is at the root of chemical evolution.

Table 7.1 presents an overview of the different classes of compounds and the corresponding biopolymers that make up the essential biogenic constituents of organisms. http://en.wikipedia.org. It is easy to enter under Natural Sciences, Biology, or Chemistry and search for different key words and chemical formulas. http://www.scirus.com/srsapp/.

Table 7.1 Classes of biogenic molecules and corresponding biopolymers with chiral carbon atoms[a]

1. **L-α-Amino acids** → *Peptides* → **Proteins** + ptm[b] → *Modified proteins*
2. *Aldehydes, ketones* → **Carbohydrates** + ptm → *Modified carbohydrates*
3. **Purine and pyrimidine bases** + *D-riboses* → *D-β-ribonucleosides + phosphate* → *D-β-ribonucleotides* → **RNA** → *Modified RNA*

4a. Glycerol[c] + ATP → di-hydroxy-acetone-phosphate → *D-glycerol-3-phosphate* → **Lipids** of Archaea, Bacteria, Eukaryotes

4b. Glycerol[c] → di-hydroxy-acetone-phosphate + NADH → *D-glycerol-1-phosphate* → **Lipids** of Archaea

4c. Glycerol[c] + fatty acids → glycerides + phosphate → *Phosphoglyceride* → **Lipid bilayer** of Bacteria, Eukaryotes

4d. Glycerol[c] + isoprene-di-phosphate → *glycerol-1-phosphate-di-ether-prenyl*-chain[d] → **Lipid monolayer** of Archaea

4e. Glycerol[c] + isoprene-di-phosphate → *di-glycerol-1-phosphate-tetra-ether-bi-prenyl* chain → **Lipid monolayer** of Archaea

[a]Main classes are in *bold* face type; molecules containing chiral carbon atoms are in *Italics*

[b]*ptm* posttranslational modifications

[c]Glycerol is a simple triose carbohydrate. In meteorites, imported glycerol and some alpha-methylated amino acids were found with an enantiomeric excess

[d]For the structure of prenyl chains, see Fig. 6.2, row 2

Part II
Scenarios of Ancient Worlds

The following scenarios took place before cells existed.

Chapter 8
The RNA World: RNA as a First Genetic System

The RNA world hypothesis proposes that RNA-based living systems predated current DNA-based life. Individual RNAs, which can store information like DNA and furthermore catalyze reactions virtually like protein enzymes, may have supported cellular and/or precellular life. The naturally occurring catalytic RNA molecules are highly specific such as the small subunit bacterial ribosomal RNA that is active as a RNA peptidyl transferase. In general, ribozymes seem to be relics from the RNA world. Many different synthetic ribozymes produced in the laboratory have diverse but defined specificities. The specificity of ribozymes is due to the intrinsic chiral centers of the RNA molecules of secondary and tertiary structure. The main evidence for an RNA world is the three essential RNA classes that act in concert to direct the specific and successive biosynthesis of proteins: tRNA, rRNA, and mRNA.

A number of essential coenzymes, ATP, CoA, NAD, and many others that consist of ribonucleotide parts and other groups, seem to be remnants of an ancient RNA world, see also Chap. 4.

Active ribozymes would effortlessly account for the origin of autocatalytic replication with feedback loops correlated with a storage and transfer of genetic information to the next generation of daughter molecules. Nucleoside-phosphates and glycerol-phosphates could not form in any primordial soup. Distinct ribozyme-catalyzed reactions, however, can account for it. The recently discovered silencing and micro-RNAs that serve to regulate gene expression might also be sort of molecular fossils. These short, double-stranded RNAs are 18–25 nucleotides long. Their role in gene expression is currently in the focus of research.

The estimated age of transfer RNA is about 3.8 billion years. There is evidence that the Genetic Code originated in the RNA world. The beauty of the world of RNA lies in the simplicity with which many different aspects of early evolution can be rationalized.

Another convincing argument for an RNA world is that today the four natural 5′-deoxy-ribonucleotides form from the four corresponding 5′-ribonucleotides by complex enzyme-mediated reactions with radicals as intermediates.

Was RNA the first genetic system? Perhaps, and likely so since other genetic systems do not supersede RNA as information carrier with the exception of DNA. Actually, polymers of diverse nucleic acid analogs based on carbohydrates, nucleobases, and

R.M. Flügel, *Chirality and Life*, DOI 10.1007/978-3-642-16977-9_8,

backbones did not show any properties that made them superior to either RNA or DNA. This holds for informational base pairing. The different backbones replaced the phosphodiester bonds by glycerol or peptide bonds. These polymers are different from the RNA peptide world discussed below.

It is difficult to explain how macromolecular RNAs emerged from the small molecules mentioned before. Perhaps a mutual catalysis in a prebiotic, relatively simple network initiated a progression of advanced stages characterized by more effective catalysis with ribozymes.

It is unresolved how far an RNA world by itself would evolve based on ribozymes alone. It seems that many active ribozymes evolved. However, some protein enzymes were genuine new inventions. For instance, the ribonucleotide reductases that use a mechanism based on radicals to convert ribonucleotides into deoxy ribonucleotides.

It is unknown when and how cooperation with amino acids, peptides, and proteins started to evolve into an RNA-protein world. However, there is an upper size limit of RNAs, which is due to a threshold error of RNA replication. The heart of the core necessary to launch the process of chemical evolution towards the RNA world must have consisted of a number of pathways for the synthesis of organic molecules from CO_2, N_2, and H_2. Additional pathways for the synthesis of amino acids, ribose, purines, pyrimidines, coenzymes, and lipids likely combined into this core. Overall, the number of pathways required to generate nucleotides is relatively small. Pyruvate, ammonia, carbon dioxide, ATP, and glyoxalate suffice to synthesize virtually the compounds required for metabolic cycles. It seems likely that once the RNA world existed that thereafter an RNA-Peptide world developed. Details are on the following website: http://www.sciencedirect.com – Cell, Volume136, Issue 4, page 599, and a description follow below.

A combination of the RNA world with the plugged pore systems recently found solves the problem of concentrating the building blocks of the nucleic acids and the nucleic acids themselves to extremely high concentrations in quasi-natural microenvironments and opens new routes to forming them. For instance, it is feasible to attempt to carry out polymer chain reactions in the plugged pores.

http://www.biophysik.physik.uni-muenchen.de/Braun/talk_pdfs/.

This website presents details of the plugged pore system.

Chapter 9
Peptide and Protein Worlds

The question of which world was first RNA or protein is not the real issue as both likely co-evolved. Thus, opportunity existed for their mutual interaction, leading to a higher level of organization and function.

It is straightforward to assume that one of the simple amino acids bound to, and put a handle on, a short RNA. Later on, peptides took over as binding partners of RNAs. The substrate-assisted catalysis used by the ribosome is a prime example for the rate acceleration of peptide bond formation. The $2'$-OH of the $3'$ terminal "A" residue of the tRNA substrate accelerates the reaction by a factor of 10^6. Strictly speaking, this scenario corresponds to a combination of two different worlds.

It seems unlikely that a protein world only existed in the early phases, however, in a pre-RNA world containing prebiotic soups with all the products of the Oro, Butlerov, Strecker, and Miller syntheses likely took place.

R.M. Flügel, *Chirality and Life*, DOI 10.1007/978-3-642-16977-9_9,

Chapter 10
Combination of Ribozymes with Peptides: The Evolving RNA Peptide World

It seems obvious that the scenarios described so far fail to explain the origin of cells and the existence of viruses. Therefore, to advance a step further toward these goals, one assumes that the aforementioned scenarios had to overlap with each other. This is feasible when the different niches that each contained one of the world scenarios had the opportunity to interact with each other. It is likely that Nature played with all the possible combinations. The difficulty here is to decide which one was first and which one followed.

RNA self-replication and ribozyme catalysts exploited peptides of random sequence and mixed chirality in their environments. These peptides were constituents of the RNA-peptide world. The next step involved the reproducible synthesis of useful, nonencoded peptides. This is analogous to the action of modern enzymes such as D-Ala-D-Ala ligase involved in bacterial cell and peptidoglycan synthesis.

RNase P cleaves a specific phosphodiester bond of tRNA precursors to form the mature 5′-end of tRNA. Bacterial RNase P is a natural occurring ribozyme composed of a catalytic RNA subunit and one or more proteins. Each bacterial RNAse P holoenzyme also consists of a small basic protein that stabilizes the folded structure of the RNA under physiological conditions. It may also help to discriminate pre-tRNA substrate from tRNA product and to mediate RNAse P dimerization. The three domains of life bacteria, Archaea, and Eukarya have similar structurally related RNAse P RNAs but different protein subunits; this indicates that RNAse P is very ancient.

Group I introns are ribozymes that carry out self-splicing in vivo. Protein components help to stabilize the folded intron structure as shown by X-ray crystal structure analysis. The protein does not have to be long; short peptides are sufficient for stabilization. Moreover, RNA templates are capable of ligating short peptides into longer active molecules. It is essential to realize that even short ones can exhibit a defined secondary structure.

Group II introns are different from group I introns in that they have an RNA secondary structure and a distinct splicing mechanism. The spliceosome, the five-RNA multiprotein complex, catalyzes mRNA splicing in Eukaryotes. It is likely that the spliceosome evolved from the group II self-splicing introns that occur in Bacteria, Archaea, and the organelles of Eukaryotes. The mechanistic and structural details are very similar. Table 10.1 compiles the known naturally occurring ribozymes.

R.M. Flügel, *Chirality and Life*, DOI 10.1007/978-3-642-16977-9_10,

Table 10.1 Naturally occurring ribozymes

Name	Activity	Remarks
23s rRNA	Peptidyl transferase	Ubiquitous
RNase P	cleaves 5′ ends of pre-tRNA	Ubiquitous
Group I introns	Self-slicing	Ubiquitous
Group II introns	Self-splicing	Ubiquitous
Hammerhead[a]	Self-splicing	Viroids
Hairpin[a]	Self-splicing	Virosoids
HDV[a]	Self-splicing of phosphodiester	Human virus
Varkud satellite[a]	Self-splicing	Fungi[b]

[a]The ribozymes function in replication by cleaving intermediates generated during rolling circle DNA replication. Reaction products are 2′-3′-cyclophosphates and 5′-OH termini
[b]The Varkud satellites were found in fungal mitochondria

It is an impressive list since several hundred different natural group I and group II introns have been shown to exist. In addition, many artificial active ribozymes have been prepared.

It is intriguing that the aforementioned natural ribozyme, the peptidyltransferase activity of the ribosome is an ancient RNA machine. Many ribosomal proteins are evolutionary conserved and contain multiple examples of glycine-, lysine-, and arginine-rich protein extensions that penetrate deep into the ribosomal RNA. Random condensation from amino acids likely produced these basic peptides.

The signal recognition particle (SRP) binds to the signal sequence of secreted proteins and directs them to the endoplasmatic reticulum of Eukaryotes or to the bacterial plasma membrane. SRP is an ancient RNA P enzyme; a portion of the RNA and one of the three subunits of the GTPase is conserved in the three domains of life. It is because of the fact that the conserved RNA can span long distances in order to impart a special sort of conformational flexibility. Thereby it bends at internal loops and bulges between helices and binds to the ribosome by RNA–RNA interactions. It is intriguing that the SRP of plants contains a homolog of the GTPase but not RNA. This indicates the transition from the RNP world to the world of protein enzymes.

The telomerase adds short telomeric DNA repeats to chromosome ends. It is composed of essential RNA and protein subunits. Telomerase is a reverse transcriptase, and furthermore, as a true RNA P enzyme, it catalyzes the synthesis of its own internal RNA template that depends on an intimate cooperation between proteins and RNA.

In the resulting Peptide-RNA World, the crucial reaction was the acylation of either one of the 2′- or 3′-OH groups of D-ribose, with the carboxyl group of an L-amino acid catalyzed by a corresponding ribozyme. The ensuing formation of an ester linkage was the first peptide bond formation. This reaction repeated itself for the generation of L-peptides. At one phase, autonomous virus-like genetic elements overlapped with a microenvironment of the Peptide-RNA World, membranes played a decisive role in forming protocells. Although artificial vesicles serve as models of protomembranes and distinct replicative reaction steps are feasible in their inside, it remained an enigma how any real protocell developed. In any case, the combination of the different scenarios opened up a new avenue to solve this conundrum.

After peptizyme and ribozyme joined forces, a prototranslation system was born!

Thus, we understand the first steps of protein biosynthesis and are standing on the threshold of a beginning DNA World.

Again, one has to make plausible assumptions. It is likely that in a pre-RNA world peptides interacted in an environment that allowed plugged pores to concentrate the required molecules. Since the plugged pores likely developed in niches of hydrothermal vents and submarine geysers, one tends to add a further combination to it.

Furthermore, a triple combination of RNA, peptide RNA, and FeS world should suffice to account for the origin of simple genetic systems and a sort of protometabolism. However, the inorganic protomembranes of the plugged pores still do not explain the existence of any protocell.

Chapter 11
The Precellular World of Viruses

Viruses are the most abundant living entities on our planet. There are more than 5,000 viral genotypes in 100 L of seawater and up to 1 million species in 1 kg of marine sediment. In addition, viruses play an important role as obligate intracellular parasites, ensuring their own reproduction, thereby affecting their host cells. The discovery of Mimivirus with the largest known viral genome of 1.2 mega-base challenges conventional definition of viruses. Some viruses seem to be relics from Nature's first attempts to generate protocells. They contain either RNA or DNA as genome. Today, they are cell parasites and have no ribosomes. According to the conventional view, viruses escaped from cells. This might well be true for some viruses. However, we do not have any knowledge of any ancient virus isolates. The virus genomes we know in the databanks and labs are young. Thus, the volatility of viral genes through time makes it almost impossible to trace them down to their ancestors. Fortunately, there are a few exceptions. Bacteriophage PM2 is a virus that propagates in the double-membrane marine bacterium Pseudomonas that belongs to the gamma proteobacteria. Phage PM contains a lipid membrane bilayer membrane reflecting its antiquity and its efforts to set up a vesicle-like structure. The composition of its phospholipids is different from that of its host bacterium. There can be little doubt that the PM2 and PRD1 virus-like ancestors possessed cell-like properties with respect to membranes. Figure 11.1 schematically shows the three-dimensional structure of the PRD1 virus. The double membrane is clearly visible in the cross section (lower panel).

Many different viruses contain enzymes required for replication and transcription. Nucleocytoplasmic large DNA and poxviruses are unusual in that they harbor many more enzymes. The genes are central to virus replication and virus structure. Nevertheless, these genes are missing from cellular genomes. They are viral hallmark genes. In addition, enzymes for capping mRNAs occur in pox and other DNA viruses. They share common motifs with eukaryotic guanyltransferases and by implication a common ancestor.

Protomembranes protected the first genetic system. The protomembrane consisted of a lipid bilayer similar to that of the PM2 virus. A corresponding bilayer of an RNA phage did not survive, unfortunately (Flügel 2010).

R.M. Flügel, *Chirality and Life*, DOI 10.1007/978-3-642-16977-9_11,

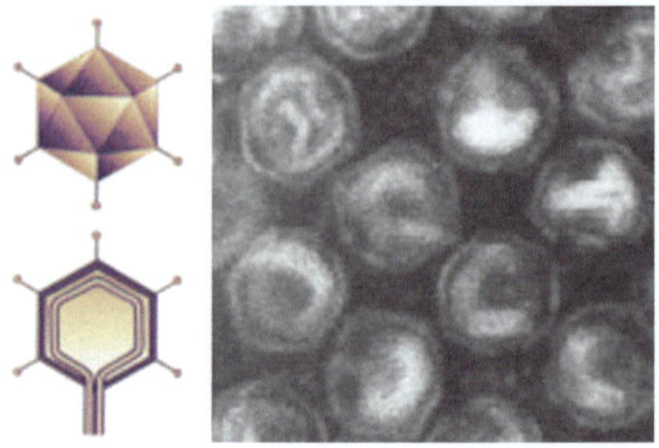

Fig. 11.1 The three-dimensional structure of the PRD virus is schematically shown on the *left hand*; the *upper* is a direct view, the *lower* part is a cross section that displays a double membrane. The electron microphotograph on the *right hand* clearly unveils a double membrane structure

Mounting evidence suggests that key components of the mitochondrial transcription and replication apparatus derive from the T-odd lineage of bacteriophages rather than from an alpha-proteobacterium as assumed by the endosymbionts hypothesis. It seems more likely that an ancestor of T-odd phages early on at the time of mitochondrial endosymbiosis acquired several mitochondrial replication genes together. The single-subunit RNA polymerase also belongs to these genes.

The Web site http://www.ncbi.nlm.nih.gov/ICTVdb/fr-fst-g.htm offers color images of the experimentally determined three-dimensional structures of many different viruses. The pictures are of great quality. In addition, the Web site offers an excellent introduction to the topology of icosahedra and triangulation numbers.

The genome of the Mimivirus, a virus from amoebas, is so large that it is comparable in size to that of a bacterium. They even have parasitic viruses that feed on them.

Chapter 12
Combinations of Different Worlds

The origin of chirality of biomolecules may have involved the potentially achiral nucleobase cytosine. Spontaneously formed chiral crystals of achiral cytosine by mechanical stirring might have been responsible for the origin of chiral biomolecules with subsequent amplification of autocatalysis.

The final and intriguing observation is that two crucial mechanisms during the origin of the molecular beginnings are autocatalytic.

First, asymmetric autocatalysis served to form chiral biomolecules. This reaction is a template-directed, self-replicating system, which successfully maintains ideal exponential growth kinetics and therefore high autocatalytic efficiency over many turnovers.

Second, autocatalytic replication of nucleic acids led to selection. Thus, life apparently emerged from inorganic nonliving matter.

R.M. Flügel, *Chirality and Life*, DOI 10.1007/978-3-642-16977-9_12,

Chapter 13
The Iron Sulfide World of Protometabolism

The FeS world is a controversial scenario. It might have started in a microenvironment or niche different from that of the RNA and protein worlds. For such a niche, hydrothermal vents or submarine geysers serve as a model that has survived till date.

The basis for the iron sulfide system as a first and archaic version of a primordial forerunner of a metabolic pathway is a chemical reaction. Under standard conditions, FeS and H_2S from submarine geochemical sources provide reducing power of about −620 mV by forming pyrite FeS_2.

$$FeS + H_2S \rightarrow FeS_2 + H_2$$

The reaction is thermodynamically favorable at higher temperatures of hydrothermal vents. Fixation of CO_2 couples with this reaction. Pyrite formation drove archaic reduction, generating a primitive form of the autocatalytic reverse citric acid cycle. The surface of pyrite minerals turned out to be playing a crucial role in the synthesis of biomolecules. Organic compounds containing energy-rich thioesters as intermediates formed. Thus, surface reactions on minerals played a crucial role in the FeS world (Wächtershäuser 2006).

On the other hand, the FeS world cannot explain the origin of chirality or the formation of glycerol. A way out of this dilemma is a combination with the consequences of meteoritic impacts on Earth as discussed before.

The presence of fixation of CO by minerals led to acetate, pyruvate, acetaldehyde, glycerol-aldehyde, acetone, methyl mercaptan (also called methanethiol), methyl-thioacetate, glycine, and aspartic acid. Alanine forms by reduction of pyruvate with ammonia. Iron sulfide and hydrogen sulfide reduce molecular nitrogen to ammonia according to the overall reaction:

$$N_2 + 3FeS + 3H_2S \rightarrow 3FeS_2 + 2NH_3$$

The redox system FeS–H_2S/FeS_2 is able to reduce dissolved molecular nitrogen to ammonia since the reaction is thermodynamically favorable with a potential of about $E_0 = -600$ mV at pH 6.5. The specific features of the catalytically active surface of fresh iron sulfide are crucial for the fixation of N_2. The surface of freshly

R.M. Flügel, *Chirality and Life*, DOI 10.1007/978-3-642-16977-9_13,

precipitated iron sulfide can be considered as a collection of [Fe–S] clusters and is comparable with those of modern enzymes that contain similar [Fe–S] clusters. Molecular nitrogen is absorbed at the surface and subsequently protonated in a stepwise manner under acidic conditions. Photos obtained by scanning electron microscopy revealed a rugged surface of iron sulfide. The mechanism of this reaction should be quite similar to that of the Haber-Bosch of ammonia as studied by Ertl, for which he got a Nobel Prize. The reaction takes about 2 weeks but is entirely consistent with and supports a chemoautotrophic origin of living systems.

1. Ni ions catalyze the following formation of carbon-oxy-sulfide

$$CO + H_2S \rightarrow COS + 2H^+ + 2e^-$$

2. Formation of methyl-mercaptan from CO_2 with FeS/H_2S

$$CO + 3FeS + 4H_2S \rightarrow CH_3SH + 3FeS_2 + H_2S + 2H_2O$$

3. Formation of methyl-mercaptan from CO and H_2S on NiS

$$CO + H_2S + 4H^+ + 4e^- \rightarrow CH_3SH + H_2O.$$

The TCA cycle is a unique anabolic central pathway for all organisms. This metabolic cycle runs reductively in several chemolithoautotrophs. The reducing chemistry of the reductive TCA cycle is the best candidate for the first protometabolic cyclic pathway in at least some environments on the early Earth. Surface reactions of CO with H_2S activated amino acids. Finally, the activated amino acids converted into peptides. Pyruvate and alkyl thiolate formed from CO and water at high pressure and temperatures of 250°C in a simulated hydrothermal microenvironment. Freshly prepared carbonylated FeS compounds are essential for a successful outcome. These experiments were and are serially connected and directed. A feedback loop convincingly correlates these reactions to the starting compounds. These results are more convincing and in strong contrast to the undirected approach of the Urey–Miller chemistry that only produced a wild mixture of many compounds. In contrast, the chemoautotrophic origin of life is a local affair. The question remains whether a metabolic cycle can evolve into a lifelike complexity. In case of the TCA cycle, this seems implausible.

Volcanic gases contain volatile P_4O_{10}, which hydrolyzes through intermediate polyphosphates and pyrophosphate that act as phosphorylation agents. These are the important sources for phosphorous. One has to bear in mind that the synthesis of phosphoenol pyruvate and D-ribose and its natural phospho-derivatives remain an unsolved puzzle in the FeS scenario.

The FeS world is a still incomplete model for building up a primitive metabolism and other pathways to enlarge the repertoire of biogenic compounds. Although limited because of the missing enantiomeric amino acids, any FeS scenario is open-ended and more attractive since it invites to combinations with other scenarios. Furthermore, the FeS world offers the advantage of focusing on minerals in plugged pore systems as important means of concentrating biomolecules in the dilute oceans of our past, molecules that are supposed to interact with each other. The reverse or reductive tricarboxylic acid cycle is a sequence of chemical reactions that are used by some bacteria to produce carbon compounds from carbon dioxide and water.

Web site: http://trstb.royalsocietypublishing.org/content/361/1474/1787.full.pdf+html

Figure 13.1 presents the TCA cyle.

The reaction is basically the citric acid cycle run in reverse. Where the Krebs cycle takes complex carbon molecules in the form of sugars and oxidizes them to CO_2 and water, the reverse cycle takes CO_2 and water to make carbon compounds. This process is used by some bacteria to synthesize carbon compounds, sometimes using hydrogen or sulfates as electron donors. The reaction is a possible candidate for prebiotic early Earth conditions and so it is of interest in the origin of life. On the early Earth, a primitive form of acetyl-CoA-like thioacetate played the role of the essential coenzyme. It has been found that some of the steps can be catalyzed by minerals. Thus, the FeS world proposes that the reverse citric acid cycle operated nonenzymatically on the primitive Earth. The question is whether it is possible to retrace other ancient metabolic pathways. Combination of the recently found plugged

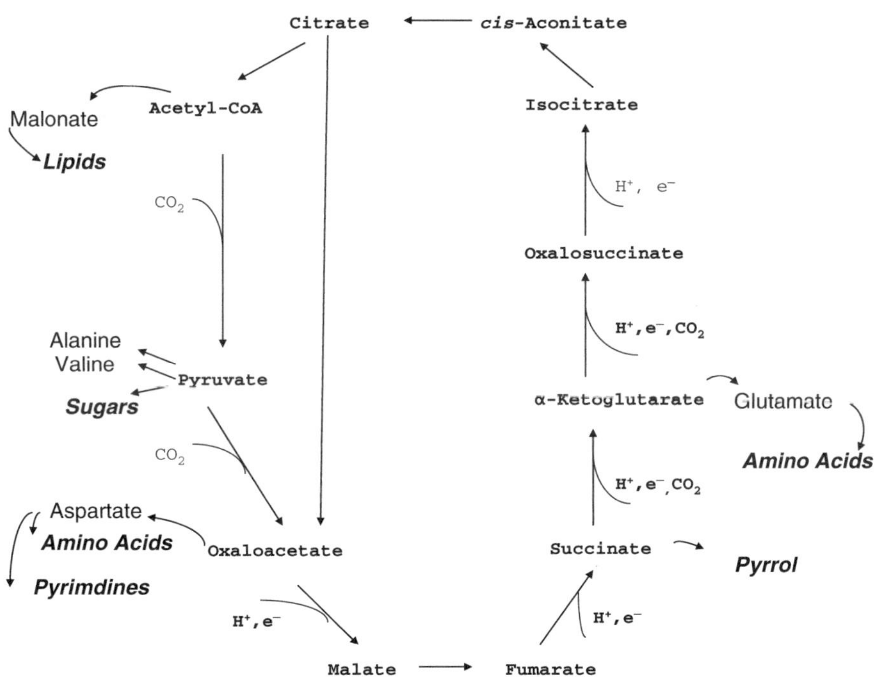

Fig. 13.1 The TCA cycle presents the main metabolic pathway for the formation and conversion of many essential molecules, such as amino acids, sugars, and lipids

pore systems with the FeS world opens new avenues to generate biopolymers. So far, the experimental evidence for mineral compartments as forerunners for protocells is insufficient.

How did the first protocells form? Recently, an excellent review was published (Meyerheinrich et al. 2010). Cell-like vesicles are sufficiently permeable to allow for the intake of charged molecules such as activated nucleotides. Subsequently, these can take part in copying templates in the protocell interior.

The three domains of living organisms have been established by the analysis of the small-subunit rRNAs. Their comparison made it possible to divide the organisms into the three well-known kingdoms (Follmann and Brownson 2009 and Woese references cited therein). Besides rRNAs, tRNAs are among the most ancient molecules, and phylogenetic trees can be constructed from them that are not inconsistent with those constructed from rRNAs and from proteins. Keep in mind that these trees are not necessarily organismal trees. Thus, a *l*ast *u*niversal *c*ommon *a*ncestor, LUCA, was introduced (Follmann and Brownson 2009 and Woese references cited therein). The notion of LUCA is based on assumptions: it is not a discrete entity. But extensive analyses have been carried out that seem to support the LUCA model.

Chapter 14
The Evolution of the Genetic Code

Many groups have attempted to trace the origin of the formation of the Genetic Code. To date, most agree that the Genetic Code was formed in the RNA world and that a primitive mechanism of translation predated the Code.

It is noteworthy that two additional amino acids were recently added to the Code, namely selenocystine, SeCys, and pyrrololysine, PyrLys. This demonstrates that the Genetic Code is still evolving. The natural expansion of the Code beyond the 20 amino acid repertoire led to the coding of modified amino acids (Ambrogelly et al. 2007).

The Code was deciphered five decades ago; some colleagues are so secretive that they do not let a word about it anymore. The Genetic Code codes for proteins in 64 triplets that consist of the four bases of RNA. A group of three adjacent letters is a codon and it codes for an amino acid. In addition, three codons stand for stop signs, that is, they determine the end of a protein chain.

But the enigma of the origin of the genetic Code stays with us. Half a dozen theories all attempt to explain the evolution of the Code.

1. The theory of "frozen accident" assumes a chance mechanism for its origin.
2. Wong's co-evolution theory postulates that codons are primarily imprints of the prebiotic pathways of the formation of amino acids that reflect faithfully the well-known enzymatic pathways that led to the biosynthesis of the amino acids. Consequently, its evolution can be elucidated on the basis of precursor–product relationships (Follmann and Brownson 2009 and Wong reference cited therein).
3. The adaption hypothesis assumes that the problem of codon assignment is a typical optimization issue with an error minimization scheme stating that near codons should encode similar amino acids. Thus, upon mutating one of the three bases of the codon, the encoded amino acid should be synonymous to the one encoded by the unmutated codon. This minimizes structural and functional deleterious effects of this mutation on the resulting protein (Di Giulio 2005).
4. The physicochemical theory considers the modern Code mainly as result of physical forces and of natural selection. This is based on the idea that chemically similar amino acids occupy close positions in the Code.
5. A stereochemical origin of the Code has been put forward that is based on the physicochemical correlation of amino acids and their codons. This idea is also called the escaped triplet theory and is supported by experimental evidence.

R.M. Flügel, *Chirality and Life*, DOI 10.1007/978-3-642-16977-9_14,

RNA aptamers artificially evolved to bind distinct amino acids that display sequence similarity with either their codon or anticodon above the level predicted by chance (Yarus 1998).

6. Recently, an asymmetric codon assignment rule was proposed. Evidence was found for a rapid early evolution of the Code via successive binary choices of 16 X_1X_2N codons. It was claimed that the other scenarios (listed above in points 1–5) could have played a role in different periods of the evolving Genetic Code.
7. A new model proposes that uses computer simulation of innovation-sharing exchange of genetic material by horizontal and communal gene transfer. Thus, an interaction between different elements took place. The postulated mechanisms account for the Code's universality, optimality, and its ability to further evolve. Similar mechanisms are required for the evolution of the translational apparatus and a corresponding Code. According to the model, protein-coding regions and gene coding for the components of translation machinery were exchanged by horizontal gene transfer. Consistent with this idea is that the mechanisms of translation are more conserved than those of replication and transcription.

Concerning the origin of cells, it was likely chaotic in the sense those large communities of mini-chromosomes – similar to transposons – were able to exchange genetic material. Generally, horizontal gene transfer played a dominant role before protocells formed. The issue of the origin of cells has been an especially frustrating and baffling problem. No solution seems to be at hand. On the contrary, it appears too complex and major gaps in our knowledge of our past are responsible for this poor situation.

Another aspect of the Code is that it was also nearly optimized for the total genetic information of coding regions, since the DNA sequences conveyed include binding sequences for regulatory and structural proteins, splicing signals, and RNA secondary structures in addition to protein-coding information. In fact, there seem to operate parallel codes for different signals (Itzkovitz 2007).

The origin of protocells was an energy-driven symbiosis between catalytically acting chiral molecules, membrane-like inorganic pore systems of submarine hydrothermal vents in liquid water, and genes combined with absorption to mineral surfaces that gained information in nonlinear mechanisms forming the basis of different scenarios. The resulting mutual interactions of the different scenarios finally combined with each other to form a protocellular system. Life is a chemical system capable of undergoing Darwinian evolution. Thus, chirality and life are connected by chiral molecules whose biochemical reactions brought about life.

References

Ambrogelly et al (2007) Natural expansion of the genetic code. Nature Chem Biol 3:29–35

Bernstein MP, Dworkin JP, Sanford SA et al (2002) Racemic amino acids from the ultraviolet photolysis of interstellar ice analogues. Nature 416:401–403

Blackmond DG, Klussmann M (2007) Spoilt for choice: assessing phase behaviour models for the evolution of homochirality. Chem Commun 3990–3996

Breslow R (1959) The mechanism of the formose reaction. Tetrahedron Lett 21:22–26

Breslow R et al (2010) Initiating prebiotic homochirality on Earth. Orig Life Evol Biosph 40:11–26

Buono FG, Blackmond DG (2003) Kinetic evidence for a tetrameric transition state in the asymmetric autocatalytic alkylation of pyrimidyl aldehydes. J Am Chem Soc 125:8978–8979

Butlerov AM (1861) Bildung einer zuckerartigen Substanz durch Synthese. Liebigs Ann Chem 120:295–298

Cintas P (2007) Sublime arguments: rethinking the generation of homochirality under prebiotic conditions. Angew Chem Int Ed 47:2918–2920

Cooper G, Kimmich N, Belisle W et al (2001) Carbonaceous meteorites as a source of sugar-related organic compounds for the early earth. Nature 414:879–883

Cordova A, Engqvist M, Ibrahem I et al (2005) Plausible origins of homochirality in the amino acid catalyzed neogenesis of carbohydrates. Chem Comm 2047–2049

Cronin JR, Pizzarello S (1997) Enantiomeric excesses in meteoritic amino acids. Science 275:951–955

Di Guilio M (1997) On the RNA world: evidence in favour of an early ribonucleopeptide. J Mol Evol 45:571–574

Di Giulio M (2005) The origin of the genetic code, theories and their relationships, a review. Biosystems 80:175–184

Di Guilio M (2008) An extension of the coevolution theory of the origin of the genetic code. Biol Direct 80:1–21

Engel MH, Macko SA (1997) Isotopic evidence for extraterrestrial non-racemic amino acids in the Murchison meteorite. Nature 389:265–267

Eschenmoser A, Loewenthal E (1992) Chemistry of potentially prebiological natural products. Chem Soc Rev 1:1–15

Ferris JP, Hagan J (1984) HCN and chemical evolution: the possible role of cyano compounds in prebiotic synthesis. Tetrahedron 40:1093–1125

Flores JJ, Bonner WA, Massey GA (1977) Asymmetric photolysis of (RS)-leucine with circularly polarized ultraviolet light. J Am Chem Soc 99:3622–3626

Flügel RM (2010) The precellular scenario of genovirions. Virus Genes 40:151–154

Follmann H, Brownson C (2009) Darwin's warm little pond revisited: from molecules to the origin of life. Naturwissenschaften 96:1265–1292

Itzkovitz SA (2007) The genetic code is nearly optimal for allowing additional information within protein-coding sequences. Genome Res 17:405–410

Joyce GF (2002) The antiquity of the RNA-based evolution. Nature 418:214–221

R.M. Flügel, *Chirality and Life*, DOI 10.1007/978-3-642-16977-9,

Kawasaki T, Sato M, Ishiguro S et al (2005) Enantioselective synthesis of near enantiopure compound by asymmetric autocatalysis triggered by asymmetric photolysis with circularly polarized light. J Am Chem Soc 127:3274–3275

Kawasaki T, Suzuki K, Hakoda Y et al (2008) Achiral nucleobase cytosine acts as an origin of homochirality of biomolecules in conjunction with asymmetric autoctalysis. Angew Chem Int Ed 47:496–499

Klussmann M, Iwamura H, Matthew SP et al (2006) Thermodynamic control of asymmetric amplification in amino acid catalysis. Nature 441:621–623

Kofoed J, Machuqueiro M, Reymond J-L et al (2004) Zinc-proline catalyzed pathway for the formation of sugars. Chem Commun 1540–1541

Kolb V, Orgel LE (1996) Phosphorylation of glyceric acid in aqueous solution using trimetaphosphate. Orig Life Evol Biosph 26:7–13

Levine MS, Kenesky CS, Mazori D et al (2008) Enantioselective synthesis and enantiomeric amplification of amino acids under prebiotic conditions. Org Lett 10:2433–2436

Melchiorre P, Carlone A, Bartoli G et al (2008) Asymmetric amino-catalysis-Gold Rush in Organic Chemistry. Angew Chem Int Ed Engl 47:6138–6171

Meyerheinrich UJ, Nahon L, Alcaraz C (2005) Asymmetric vacuum UV photolysis of the amino acid leucine in the solid state. Angew Chem Int Ed Engl 44:2–5

Meyerheinrich UJ, Filippi J-J, Meinert C et al (2010) On the origin of primitive cells. Angew Chem Int Ed Engl 49:3738–3750

Miller SL (1953) A production of amino acids under possible primitive earth conditions. Science 117:528–529

Morowitz HJ (1969) A mechanism for the amplification of fluctuations in racemic mixtures. J Theor Biol 25:491–494

Munoz-Caro GM, Meyerheinrich UJ, Schutte WA et al (2002) Amino acids from ultraviolet irradiation of interstellar ice analogues. Nature 416:403–406

Pereto J et al (2004) Ancestral lipid biosynthesis and early membrane evolution. Trends Biochem Sci 29:469–477

Pizzarello S (2006) The chemistry of life's origin: a carbonaceous meteorite perspective. Acc Chem Res 39:231–237

Pizzarello S, Weber AL (2004) Prebiotic amino acids as asymmetric catalysts. Science 303: 1151–1153

Pizzarello S, Weber AL (2010) Stereoselective syntheses of pentose sugars under realistic prebiotic conditions. Orig Life Evol Biosph 40:3–10

Powner MW, Garland B, Sutherland JD (2009) Synthesis of activated pyrimidine ribonucleotides in prebiotically plausible conditions. Nature 459:239–242

Scumlimbrene BR, Miller SJ (2001) Discovery of a catalytic asymmetric phosphorylation through selection of a minimal kinase mimic: a concise total synthesis of D-myo-inositol-phosphate. J Am Chem Soc 123:10125–10126

Shibata T, Morioka H, Hayase T, Choji K, Soai K (1996) Highly enantioselective catalytic asymmetric auto-multiplication of chiral pyrimidyl alcohol. J Am Chem Soc 118:471–472

Soai K, Kawasaki T (2006) Discovery of asymmetric autocatalysis with amplification of chirality and its implication in chiral homogeneity of biomolecules. Chirality 18:469–478

Szostak JW (2010) On the origins of primitive cells: from nutrient intake to elongation encapsulated nucleotides. Angew Chem 49:3738–3750

Szostak JW et al (2001) Synthesizing life. Nature 409:387–390

Viedma C et al (2008) Evolution of solid phase homochirality for a proteinogenic amino acid. J Am Chem Soc 130:15724–15725

Wächtershäuser G (2000) Living on a ligand sphere. Science 289:1307–1308

Wächtershäuser G (2006) from volcanic origins of chemoautotrophic life to Bacteria, Archaea and Eukarya. Phil Trans Soc B 361:1787–1808

Yarus M (1998) Origins of the genetic code: the escaped triplet theory. Annu Rev Biochem 74:179–198

Subject Index

(Numbers in ***italics*** indicate Figures; numbers underlined refer to Table).

MIX
Papier aus verantwortungsvollen Quellen
Paper from responsible sources
FSC® C105338

If you have any concerns about our products,
you can contact us on
ProductSafety@springernature.com

In case Publisher is established outside the EU,
the EU authorized representative is:
Springer Nature Customer Service Center GmbH
Europaplatz 3, 69115 Heidelberg, Germany

Printed by Libri Plureos GmbH
in Hamburg, Germany